General Radiography

Medical Imaging in Practice

Series Editors: Christopher Hayre, Assistant Professor in Medical Imaging
Institute of Applied Technology, UAE

For more information about this series, please visit: https://www.crcpress.com/
Medical-Imaging-in-Practice/book-series/MIIP

General Radiography

Principles & Practices

Edited by

Christopher M. Hayre
William A. S. Cox

CRC Press
Taylor & Francis Group
Boca Raton London New York

CRC Press is an imprint of the
Taylor & Francis Group, an **informa** business

First edition published 2020
by CRC Press
6000 Broken Sound Parkway NW, Suite 300, Boca Raton, FL 33487-2742

and by CRC Press
2 Park Square, Milton Park, Abingdon, Oxon, OX14 4RN

© 2020 Taylor & Francis Group, LLC

CRC Press is an imprint of Taylor & Francis Group, LLC

International Standard Book Number-13: 978-0-367-14987-1 (Hardback)
International Standard Book Number-13: 978-0-367-49767-5 (Paperback)
International Standard Book Number-13: 978-1-003-04727-8 (eBook)

Library of Congress Cataloging-in-Publication Data

Names: Hayre, Christopher M., editor. | Cox, William A. S., editor.
Title: General radiography : principles & practices / edited by Christopher M. Hayre and William A.S. Cox.
Description: First edition. | Boca Raton : CRC Press, 2020. | Includes bibliographical references and index. | Summary: "This book explores pertinent themes related to the practice of general radiography within the medical imaging setting. It focuses on important concepts linked to image acquisition, dose optimization, patient care and discussions surrounding learning and teaching within the academic environment. We have found that the very nature of general radiography sparks an array of clinical discussion and provides numerous opportunities for future research. Through the experiences of practitioners and academics transnationally this book offers readers the ability to learn, reflect and consider their own practice"– Provided by publisher.
Identifiers: LCCN 2020004492 (print) | LCCN 2020004493 (ebook) | ISBN 9780367149871 (hardback) | ISBN 9781003047278 (ebook)
Subjects: MESH: Radiography
Classification: LCC RC78.A1 (print) | LCC RC78.A1 (ebook) | NLM WN 200 | DDC 616.07/572–dc23
LC record available at https://lccn.loc.gov/2020004492
LC ebook record available at https://lccn.loc.gov/2020004493

Dedication

Dr. Christopher Hayre would like to dedicate this book to his family, Charlotte, Ayva, Evelynn, and Ellena. Love to all. William Cox would like to dedicate this book to his family at home and abroad.

Contents

SECTION 4   Learning, Teaching, and Education   193

Acknowledgments

The editors would like to thank all of the authors who have contributed to this book for their commitment, hard work, and determination. It has been a pleasure to work with you and bring together this collection of chapters. We believe that these demonstrate the continued importance of general radiography within the medical imaging environment. We have personally found this to be an exciting and enriching project, which we hope readers will enjoy and utilize.

Preface

This book explores pertinent themes related to the practice of general radiography within the medical imaging setting. It focuses on important concepts linked to image acquisition, dose optimization, patient care, and discussions surrounding learning and teaching within the academic environment. We have found that the very nature of general radiography sparks an array of clinical discussion and provides numerous opportunities for future research. Through the experiences of practitioners and academics transnationally, this book offers readers the ability to learn, reflect, and consider their own practice.

This book is divided into four sections. Section 1 provides an introductory perspective on general radiography in the contemporary setting. Section 2 provides a discussion of the importance of image acquisition and dose optimization. These two interlinked themes are central to the role of radiographers and sound patient care. Section 3 contains a selection of perspectives on patient care in a variety of settings including pediatrics, critical care, and trauma. Finally, Section 4 is concerned with learning, teaching, and education in radiography.

This text will be helpful for a range of readers. Students who are writing for undergraduate dissertations and/or studying for written exams will find the theoretical examples stimulating, thus enhancing their learning. For academics, practitioners, and prospective researchers, this book offers patient-focused themes, which remain highly transferable into lecture halls and/or via general discussion.

It is anticipated that readers will find this collection of radiographic work not only useful for informing their own professional development, but we also hope to stimulate novel discussions and arguments in order to develop the practice of general radiography internationally.

Dr. Christopher M. Hayre

William A.S. Cox

Contributors

Dr. Christopher M. Hayre
Senior Lecturer
Diagnostic Radiography
Charles Sturt University
New South Wales, Australia

Kevin McHugh
Lecturer in Diagnostic Radiography
University of Portsmouth
United Kingdom

Professor Rob Davidson
Professor of Medical Imaging
University of Canberra
Australia

Shantel Lewis
Lecturer in Medical Imaging and
    Radiation Sciences
University of Johannesburg
South Africa

Emma Hyde
Head of Diagnostic Imaging
University of Derby
United Kingdom

Professor Maryann Hardy
Professor of Radiography and
    Imaging Practice Research
University of Bradford
United Kingdom

Dr. Chandra Makanjee
Department of Medical Radiation
    Sciences

School of Clinical Sciences
University of Canberra
Australia

Tom Campbell-Adams
Department of Diagnostic
    Radiography
University of Portsmouth
United Kingdom

Allen Corrall
Specialist Paediatric
    Radiographer
University Hospital Southampton
    NHS Foundation Trust
Southampton, United Kingdom

Naomi Shiner
Senior Lecturer in Diagnostic
    Imaging
University of Derby
United Kingdom

Dr. Iain MacDonald
Senior Lecturer in Diagnostic
    Radiography
Institute of Health
University of Cumbria

Julie Hendry
Associate Professor and Associate Dean
Kingston and St George's Joint
    Faculty, Health, Social Care and
    Education
Department of Radiography
University of London

Section 1

Introductory Perspective

1 General Radiography in the Contemporary Setting

Christopher M. Hayre

SETTING THE SCENE

The practice of general radiography requires sound knowledge and understanding of radiographic principles, coincided with good interpersonal collaboration and person-centeredness. Looking back, I often reflect on my first experiences within this imaging modality as a radiography student, recollecting how central this imaging modality was at treating and managing a wide range of suspected health conditions within the X-ray environment. Further, I look back on the transition from film-screen (FS) to computed radiography (CR), and then digital radiography (DR). This technological change led to my own PhD study, later uncovering the effects of advancing technology and how it affected practitioners and patients alike.

General radiography, then (as the title of this book suggests) requires a more 'generalist' set of skills in order to undertake a wide range of projection examinations. It also hints at the possibility of general imaging being regarded as 'non-specialist' by convention, and thus arguably inferior when compared to other 'specialist imaging roles'. This argument is supported by reflecting on existing job opportunities worldwide, whereby general radiographer positions are often associated with lower salaries, positioned in lower pay bandings when compared to other imaging roles affiliated with computed tomography (CT), magnetic resonance imaging (MRI), mammography and interventional imaging, for instance. My reflections of academia are also similar. Senior lecturer positions tend to seek successful applicants with 'specialist experience' in order to meet the demands of the ever-evolving needs of higher education institutions. This book is in part a celebration and a reminder of the virtues required to perform general radiographic examinations optimally, which remain underscored by principles and practices in both clinical and academic contexts. Henceforth, up and coming chapters reaffirm the utility of general radiographic principles and practices, which is anticipated to add value to readers transnationally.

The book begins by outlining pertinent theory regarding image acquisition and dose optimization (Section 2). The book then draws from experienced practitioners by recognizing the value of interprofessional working among radiographers within the hospital environment, and more importantly, reminds us of sustaining

3

person-centeredness in trauma and critical care scenarios (Section 3). Finally, the editors have selected chapters concerning learning, teaching, and education in diagnostic radiography, and how virtual reality and other innovative forms of pedagogical and andragogical methods are being considered in order to enhance the delivery and experiences of undergraduate radiography students education (Section 4).

It is anticipated this book will be of use to a number of audiences. First, for student radiographers, sections 2 and 3 not only offer foundational principles but also provide experiences of delivering patient care and sound interprofessional working. Second, academics will find the content useful for knowledge transfer, while also utilizing topics for discussion. Finally, it is anticipated that postgraduate students and early career researchers will find the content and references helpful in order to define their own research objectives, and importantly uncover their own areas of original research.

In addition to the aforementioned, this book reaffirms the value of general radiography as it continues to remain the most frequent imaging modality performed worldwide, thus a useful criterion for assessing our professionalism and competency. This also raises two additional areas for discussion. First, because general radiography continues to constitute a large frequency of medical imaging examinations worldwide, it is argued that it remains important for the radiographic community to continuously challenge and reflect upon its clinical application(s). Second, while it is proffered that general radiography is perhaps seen inferior or 'easier' than say a 'specialist modality', this does not negate our need for practicing it optimally. From the author's own observation(s), he was exposed to experimental research and sound pedagogy seeking to minimize ionizing radiation for general imaging examinations as an undergraduate student, yet an immediate disconnect was observed between classroom discussions and what was practiced clinically. For example, the use of appropriate exposure factors (Hayre, 2016) selection through phantom work, and/or the application of lead (Pb) rubber are two examples of such dichotomous practices (Hayre et al., 2017). Further, questions in the author's own work have raised questions over person-centeredness (Hayre, Blackman and Eyden, 2016) and/or whether radiographers in the past have had the appropriate knowledge and understanding to acquire digital radiographs optimally (Hayre et al., 2017). These questions, and then subsequent findings, were not to pick holes, but to question, challenge, and remind us of our need to ensure sound radiographic competencies within an imaging modality that continues to represent the largest portion of what we do as diagnostic radiographers. In short, and regardless of technological advances an array of developmental opportunities exists as diagnostic radiographers.

GENERAL RADIOGRAPHY: A PLATFORM FOR INNOVATION AND CHANGE?

The general radiography environment has historically offered a platform for both innovation and change. It is generally accepted that to practice general radiography 'well' requires a competent practitioner who can produce diagnostically acceptable images while ensuring that patients are treated and cared for in an altruistic manner. Further, with increasing time pressures, radiographers also require the physical skill to manoeuver X-ray equipment and obtain radiographs in an effective, but timely manner for patients and referring clinicians. While these attributes alone often ensure good radiographic outcomes, it is argued here that both innovation and organizational change play a significant role in the general imaging environment. For example, general radiography first offered abnormality detection opportunities for radiographers, commonly referred to as the 'red dot' system. This enabled radiographers to assist physicians by highlighting pathology (typically fractures) in the emergency department. In recent years, this has now evolved and led to the development of radiographer reporting offering a medico-legal diagnosis, a role commonly undertaken by radiologists. While some contention around the role development of radiographers exists in the literature (RANZCR, 2018, RCR, 2018, Hayre and Atutornu, 2019), the practice of radiographer reporting has become a recognized form of advanced practice, which has transcended into other imaging modalities and now become an organizational norm in most departments within the United Kingdom.

The example above is a reminder of the role general imaging plays in delivering both innovation and change, due to the emergence of new roles in order to enhance outcomes for patients. Another recent development in radiography has been that of advancing technology and the suggestion of a digital radiography champion (DRC) (Hayre et al., 2017). While this role is merely theoretical, it does aim to bridge both opportunities and challenges commonly associated with digital radiography, such as the potential for lowering ionizing radiation for patients, but also recognizing and educating radiographers of overexposure (ibid). As technology will continue to remain a significant driver for the radiography profession in years to come we may need to think outside the box in terms of delivering alternate forms of advanced practice in order to not only meet the operational needs of the department, but more importantly enhance the care and experiences of patients.

The rationale for a DRC in contemporary practice is grounded on the view that digital radiography offers significant dose optimization, without compromising image quality. However, there has also been a suggestion whether this evidence base is actually being transferred into the clinical environment and/or whether we are simply experiencing a 'drift' away from research evidence (Snaith, 2016), as radiographers may not always conform to the delivery of sound ionizing radiation to patients (Hayre, 2016). This offers an opportunity to bridge a theory-practice gap by means of introducing a DRC practitioner. For example, as a specialist in image acquisition and dose optimization a DRC could help connect theory and practice by undertaking and reflecting on dose audits/empirical research, deliver training and learning to staff, and conduct clinical trials in order to help enhance imaging protocols and policies with key stakeholders. Further, a DRC could become a conduit between medical physicists, referrers, radiographers, and patients by ensuring clinical questions remained radiographically sound, but also ensure that radiation doses are continuously optimized in light of technological advances.

In the past, general radiography has delivered innovative practices that have become commonplace. It is through innovation that novel opportunities can emerge, which can later lead to change and become cultural normality. For example, in the example of the DRC, this can ensure that technological advancements are optimized for patients and may allow us to rethink how 'advanced practiced' is considered. At present, increasing emphasis on 'advanced practiced' is linked to image interpretation, whereby we mimic roles performed by medical doctors. However, there is an argument that as technology continues to offer enhanced hardware and software capabilities one ongoing challenge for radiographic practitioners will be the accountability of optimizing image quality for patients, while using as little radiation as possible.

SUMMARY

This introductory chapter offers a perspective that considers general radiography as a specialism. Although general radiography could be associated as an elementary step into the radiographic profession for radiographers, it is considered here as a modality that can offer innovation and change. First, the volume of examinations undertaken, supported with the multifaceted approach in which general radiography supports the management of patients with ill-health reflects this. There remains a wealth of opportunism for research, innovation, and change in an imaging modality that has historically demonstrated such virtues. The forthcoming chapters celebrate and reaffirm this by offering theoretical, clinical, and academic discussions with an overall aim of advancing knowledge and generating further discussion within an array of academic and clinical contexts.

REFERENCES

Hayre, C.M. (2016) 'Cranking up', 'whacking up' and 'bumping up': X-ray exposures in contemporary radiographic practice. *Radiography*, 22 (2), pp. 194–198.

Hayre, C.M. and Atutornu, J. (2019) Is image interpretation a sustainable form of advanced practice in medical imaging? *Journal of Medical Imaging and Radiation Sciences*, 50 (2), pp. 345–347.

Hayre, C.M., Blackman, S., Carlton, K. and Eyden, A. (2017) Attitudes and perceptions of radiographers applying lead (Pb) protection in general radiography. *Radiography*, 24 (1), pp. e13–e18.

Hayre, C.M. Blackman, S. Eyden, A. (2016) Do general radiographic examinations resemble a person-centred environment? *Radiography*, 22 (4), e245–251.

Royal College of Radiologists. *The Radiology Crisis in Scotland: Sustainable Solutions Are Needed Now.* [Online] Available at: www.rcr.ac.uk/posts/radiology-crisis-scotland-sustainable-solutions-are-needed-now (Accessed: 07/08/2018).

Snaith, B. (2016) Evidence based radiography: Is it happening or are we experiencing practice creep and practice drift? *Radiography*, 22 (4), pp. 267–268.

The Royal Australian and New Zealand College of Radiologists. *Image Interpretation by Radiographers – Not the Right Solution.* [Online] Available at: https://www.sor.org/system/files/news_story/201205/IPAT%20Final%20Report%2012%2034%204%2028%20(2).pdf (Accessed: 7/4/2020).

Section 2

Image Acquisition and Dose Optimization

2 Digital Receptors

Kevin McHugh

DIGITAL IMAGE ACQUISITION

The launch of the now defunct National Programme for Information Technology (NPfIT) in 2002 highlighted that considerable change was required with regard to diagnostic imaging. Traditionally, images were acquired using radiographic film and the NPfIT necessitated moving from analogue to digital acquisition. This introduced the concept of Digital Radiography (DR) and technology that was able to perform the functions of radiographic film. The aim of these technologies is to convert the radiation intensities leaving the patient into electrical charge and then subsequently a binary number via an analogue to digital converter (ADC) for display and archival purposes.

To produce images in a digital format, it was necessary to discard the use of film/screen radiography (FSR) and introduce electronically readable devices that have the ability to record the latent image. In practice, these devices are known as Digital Receptors (DRs) and take the form of Computed Radiography (CR), which uses Photostimulable Phosphor Plates (PSPs) and digital Radiography (dR), which uses Flat Panel Detectors (FPDs). There are two types of FPDs available: direct digital Radiography (ddR) and indirect digital Radiography (idR). It is important to convey here that the abbreviation DR incorporates all devices that are able to record the latent image (Figure 2.1), whereas the abbreviation dR refers to the use of FPDs to acquire an image.

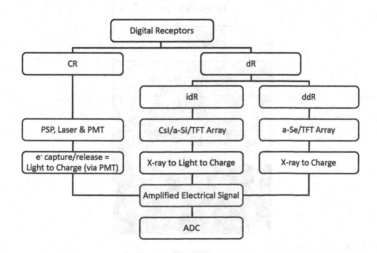

FIGURE 2.1 Schematic Overview of Digital Receptors.

The devices used to produce digital images can be crudely thought of as a receptor that is in the form of a matrix of many individual pixels. Upon exposure to radiation intensities leaving the patient, the pixels absorb X-ray photons and the acquisition device converts the X-ray photons into an electrical signal. It can be speculated that if the X-ray photons have interacted with an area that is radiolucent, then the electrical signal is significantly large. Conversely, if the X-ray photons have interacted with an area that is radiopaque, then the electrical signal is significantly small. The electrical signal that is generated is then subsequently converted to a binary number via an ADC.

COMPUTED RADIOGRAPHY

CR involves the use of an imaging plate known as a Photostimulable Phosphor Plate (PSP), which is housed within a protective cassette. Once the imaging plate has received an exposure, it is necessary to extract the PSP from the protective cassette and read it using a CR reader. Subsequently, the receptor being used here mimics the action of FSR, whereby after being exposed it needs to be processed.

The plate itself has a phosphor layer that is crystalline in structure and contains either barium fluorohalide (BaFl) or caesium bromide (CsBr), both of which are doped with europium (Eu^{2+}). The phosphor layer is coated onto a base using polymers that act as glue to hold it in place. A clear solvent is then coated over the phosphor to seal it to protect the phosphor from physical damage. A black reflective base under the phosphor helps improve image resolution by reducing dispersion of light as the CR reader processes the plate. The black base also allows for a thicker phosphor layer providing better absorption efficiency. These are all mounted onto a lead sheet that absorbs excess photons and reduces backscatter whilst an aluminum panel is used as a backing layer (Figure 2.2).

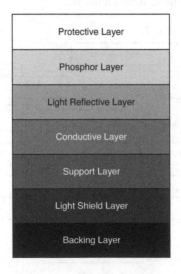

FIGURE 2.2   Construction of a CR Imaging Plate.

PHYSICAL PRINCIPLES OF IMAGE FORMATION FOR CR

A number of theories have been proposed to explain the physical principles of image formation for CR (Leblans *et al.*, 2011). It is interesting to note that the seminal paper concerning the physics of CR by Rowlands (2002) infers that this process is controversial and differs between specific PSPs. Many academic sources which refer to digital imaging (Carlton & Adler, 2013; Carter & Vealé, 2008; Holmes *et al.*, 2014; Lança & Silva, 2013; Oakley, 2003) provide varied detail in regard to latent image formation for CR. They all refer to electron excitation and capture but proffer differing explanations in respect to ionization and the liberation of an excited electron (ibid), in addition to the exact process of electron capture (Carter & Vealé, 2008; Lança & Silva, 2013; Oakley, 2003). The principle of latent image formation using this type of device is based upon electron band theory.

ELECTRON BAND THEORY

When two atoms join to become a molecule, the orbital shells (which are said to be quantized energy states) fuse together to create molecular orbits (Figure 2.3).

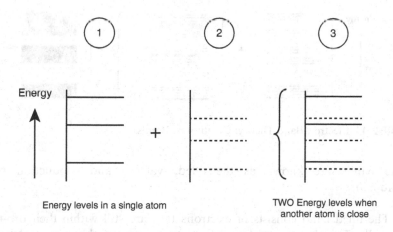

FIGURE 2.3   Electron Band Theory.

This is primarily due to electrons in orbital shells becoming influenced by electric fields of other atoms. Consequently, this creates energy levels and as more atoms start to fuse, these energy levels can become continuous therefore creating bands (Figure 2.4).

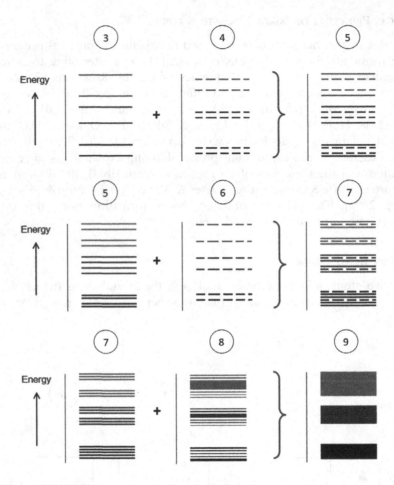

FIGURE 2.4 Electron Band Theory: Creation of Bands.

The bands are known as the filled, valence and conduction bands (Figure 2.5).

- The filled band consists of electrons that are still within their orbital shells. The electrons are considered to be part of the atom and have a strong attraction to the nucleus.
- The valence band consists of electrons that are still considered part of the atom. Electrons within this band are considered to have a high potential energy to remain in orbit as they are usually in orbital shells that are further away from the nucleus.
- The conduction band is an energy level above the valence band and any electrons in this band are said to be free electrons and can move freely within the material. If this occurs, the material is said to demonstrate conductive properties.

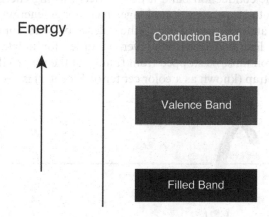

FIGURE 2.5   Electron Band Theory: Formation of Bands.

The gap between the conduction band and the valence band is known as the forbidden gap or zone. No electrons exist here but the size of the gap will dictate the conductive ability of the material. With conductors, the conduction band and valence band tend to overlap but some materials where a gap exists can become conductors. This occurs if electrons can be excited from the valence band to the conduction band. Any material where it is possible to do this is known as a semiconductor.

LATENT IMAGE FORMATION IN CR

The phosphor layer of the CR plate is considered similar to a semiconductor. Upon exposure to radiation, the europium within this layer ($BaFl:Eu^{2+}$ or $CsBr: Eu^{2+}$) is ionized from Eu^{2+} to Eu^{3+}, therefore liberating an electron (Fujifilm Imaging Plate Manual, n.d.; Rowlands, 2002; Seeram, 2019). The 'excited' electron residing in the valence band now has enough energy to jump into the conduction band (Figure 2.6).

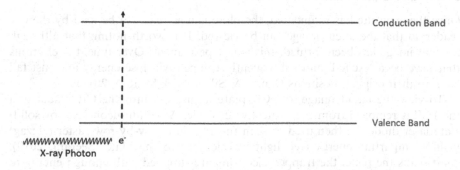

FIGURE 2.6   Computed Radiography: Electron Excitation.

Once within the conduction band it travels freely, losing energy over time. As the electron starts to lose energy, it no longer has enough energy to remain in the conduction band and wants to return to the valence band to unite with the europium ion it was liberated from. However, in order for a latent image to be formed, the electron must be stopped from falling all the way to the valence band and so an energy trap (known as a color center or F-center) is used (Figure 2.7).

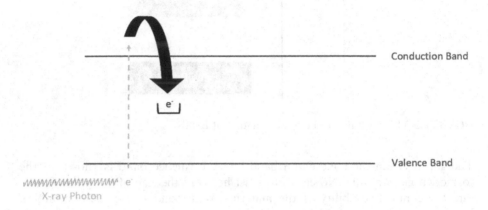

FIGURE 2.7   Computed Radiography: Electron Capture.

The addition of europium changes the structure of the crystal. Consequently, the electrons actually become trapped within the halide portion of the phosphor layer due to defects in the lattice that are present in the crystal as a result of the addition of europium (Fujifilm Imaging Plate Manual, n.d.; Leblans *et al.*, 2011). Subsequently, these defects become traps for the electrons that are leaving the conduction band. The process of electron release (ionization) and capture (energy traps) happens numerous times and is proportional to the X-ray exposure received. This contributes to the formation of the latent image.

LATENT IMAGE TO VISIBLE IMAGE IN CR

Once the exposure has terminated, the plate is now ready to be read by the CR reader so that the latent image can be viewed. It is worth noting that although a latent image has been formed, this is not permanent. Over time, the electrons that have been 'excited' and subsequently trapped will lose energy and just fall back to their original positions (Lança & Silva, 2013; Seeram, 2019).

To view the latent image, the CR plate is inserted into the CR reader and the PSP is removed from the protective cassette. A helium-neon laser or solid-state laser diode are then used to scan the PSP in a row-by-row raster (zigzag) fashion imparting energy (red light wavelength) to the plate (Seeram, 2019). As it scans the plate, the trapped electrons are supplied with enough energy to move back to the valence band and unite with the ionized europium atoms.

There are numerous theories postulated in regard to what happens to the electrons when they have been supplied with energy (Leblans *et al.*, 2011). Both the Fujifilm Imaging Plate Manual (n.d.) and Seeram (2019) suggest that the electrons are re-excited to the conduction band prior to being united with the ionized europium atoms. It could be argued here that if this was the case then the process of electron excitation and capture would re-occur to form a new latent image. The fundamental point is that the electrons that are trapped are stimulated using the laser to fall back to the valence band, which in turn reunites them with ionized europium atoms. The transformation of europium from Eu^{3+} to Eu^{2+} liberates energy (Figure 2.8), which is emitted as violet/blue light (Lança & Silva, 2013; Seeram, 2019). It is important to note here that the emission of violet/blue light is what is measured to form the resultant image, not the phosphor layer of the CR plate luminescing.

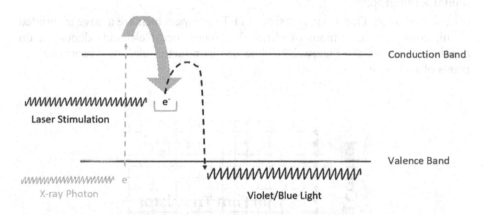

FIGURE 2.8 Computed Radiography: Electron Release.

In order to detect the light that is emitted from the transformation of europium, a filter is used to absorb the red light from the laser and allow transmission of the violet/blue light. The violet/blue light passes through a guide to a series of photomultiplier tubes (Rowlands, 2002; Seeram, 2019) or in some CR readers a number of solid-state photodiodes (Lança & Silva, 2013). The role of these components is to convert the light emitted into an electrical signal, which is amplified before passing onto an ADC. This signal is proportional to the intensity of light emitted and is converted into digital data by the ADC.

Although the laser of the reader provides energy for the trapped electrons to return to their original state, not all of the trapped electrons return (Lança & Silva, 2013). In order for the plate to be used again, these electrons must be returned to their original position. This is achieved by flooding the plate with bright white light and is known as the erase cycle. Most CR readers have a primary and secondary erase cycle. The primary cycle is used when the CR plate is being read after being exposed, whilst the secondary cycle is used for

CR plates that have not been used within 48 hours but may have been exposed to secondary radiation (Fujifilm Medical Systems CR User Guide, 2004). In both, the intense light provides the electrons with the correct amount of energy to return. This explains why once exposed, the CR plate needs to be handled under safelight conditions similar to conventional film.

The erase cycle also plays a pivotal role in erasing any remnant image. Manufacturers recommend that CR plates should be left for two minutes prior to being inserted into a CR reader. The rationale is that this time period is sufficient for all 'excited' electrons to become settled within the traps. If the CR plate is processed too soon, it is possible to re-stimulate electrons that have just become trapped or those that are falling into traps back to the conduction band. Consequently, the process of latent image formation occurs once more, creating a phenomenon known as a remnant image.

digital Radiography

dR is based upon Thin Film Transistor (TFT) arrays, which are a large integrated circuit consisting of millions of identical semiconductor elements deposited on a substrate material (Figure 2.9). The semiconductor elements represent the pixels of an image.

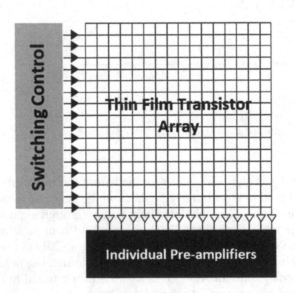

FIGURE 2.9   digital Radiography (dR) TFT Matrix.

Within medical imaging, TFT arrays tend to be used within Flat Panel Detectors (FPD's) that can be both indirect and direct in their capture and conversion of X-ray photons.

Indirect Digital Radiography (idR)

For idR, the TFT array consists of a sheet of glass covered with a thin layer of amorphous silicon. This is then integrated with arrays of photodiodes that are deposited onto thin film transistors (TFTs). The array itself is then covered with a scintillator with each photodiode being thought of as a pixel (Figure 2.10).

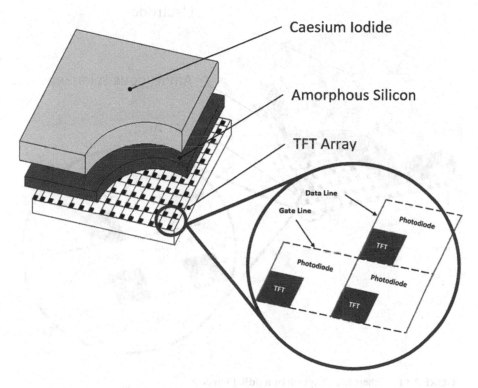

FIGURE 2.10 Schematic Diagram of an idR Detector.

The photodiodes are covered by a layer of either caesium iodide (CsI) or gadolinium oxysulfide (GdO_2S_2) that acts as a scintillator, i.e. producing light when X-rays interact with it (Lança & Silva, 2013). It is worth noting here that both CsI and GdO_2S_2 are described as scintillators rather than intensifying screens. The green visible light produced by both scintillators (Huda, 2016; Zhao et al., 2004) is absorbed by the photodiode on each TFT, which in turn produces electrons that are stored on the capacitance of the photodiode (Lança & Silva, 2013). The electrons are produced due to the layer of amorphous silicon covering the photodiode. These will form the electrical signal to be digitized.

Direct Digital Radiography (ddR)

For ddR, the TFT array consists of a sheet of glass onto which is mounted a detector element array that consists of a capacitor and a TFT. A layer of amorphous selenium is directly coated onto the detector elements with each capacitor being thought of as a pixel. An electrode then covers the detector elements (Figure 2.11).

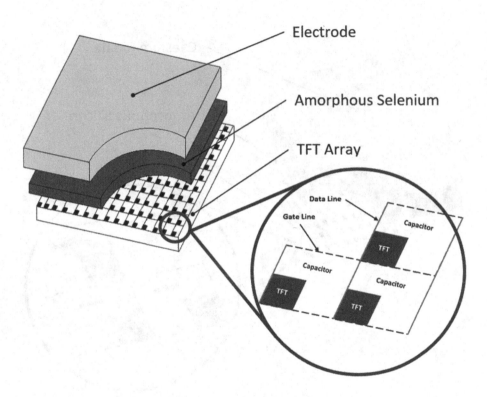

FIGURE 2.11   Schematic Diagram of a ddR Detector.

This type of receptor converts radiation intensities exiting the patient directly into an electrical signal (Lança & Silva, 2013; Seeram, 2019). The layer of amorphous selenium is used to convert X-ray photons into negative and positive charges in proportion to the level of X-ray exposure. The negative charges in this instance are electrons whereas the positive charges are holes. A voltage is then applied across the electrode that forms an electric field across the detector element array. The generation of an electric field attracts the negative charges (electrons) to the electrode thus leaving the detector element array. Conversely, the positive charges (holes) are repelled to be collected by the capacitor with minimal lateral diffusion (Lança & Silva, 2013). The collection of positive charges forms the electrical signal to be digitized.

It is worth noting that some texts (Carlton & Adler, 2013; Carroll, 2018; Holmes *et al.*, 2014; Seeram, 2019) describe and/or illustrate ddR detectors as having a secondary electrode at the TFT array called a 'charge collecting electrode' or 'pixel electrode'. The charge collecting electrode or pixel electrode is described as many minute electrodes (Holmes *et al.*, 2014; Lança & Silva, 2013). In this instance, these texts are referring to the capacitors. In addition, Carlton & Adler (2013), Carroll (2018) and Holmes *et al.* (2014) describe that it is the polarity of the electrode which separates the negative and positive charges. The electrode is certainly responsible for separating them and it can be described as having polarity when a voltage is applied through it. However, 'the electric field causes them [negative and positive charges] to move toward the TFT elements where they are collected and stored' (Seeram, 2019, page 75).

The electrical signal collected in both type of receptors is read out by using the TFT array. The TFT array is a large integrated circuit that has horizontal and vertical conductors known as gate lines and data lines, respectively. These lines create a circuit that mirrors a matrix with each detector element (photodiode/capacitor) being considered a pixel. The TFTs within the array act as switches and are connected to a switching control circuitry via the gate lines in a manner that allows all the switches in a row of the array to be operated simultaneously. The output from each pixel (photodiode/capacitor) is then connected in columns via data lines with individual pre-amplifiers (Figure 2.12).

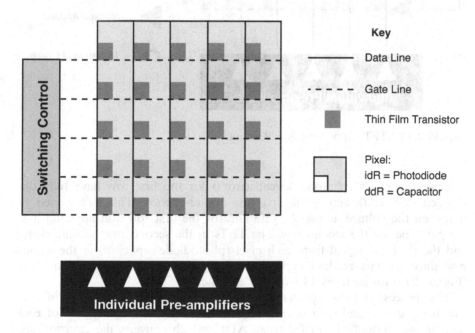

FIGURE 2.12   TFT Array.

During the readout process, the electrical signals collected are read out one row at a time and this is controlled by the switching control unit. During the X-ray exposure all the TFTs across the array are open, however as soon as the exposure is terminated the TFTs in the first row of the array are closed. This is achieved by applying a voltage across the gate line for the first row. The amount of electrical signal from each pixel (photodiode/capacitor) in the first row since the last readout cycle is then migrated to the adjacent data line (Figure 2.13) for each pixel (photodiode/capacitor).

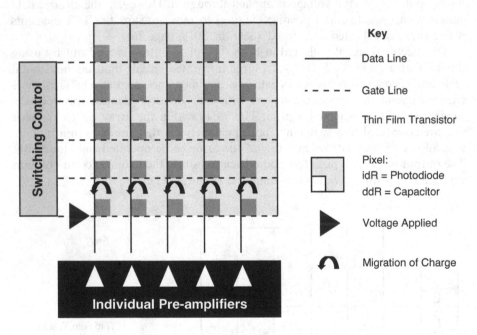

FIGURE 2.13   TFT Array: First Row Readout.

Once the pixels (photodiodes/capacitors) for the first row have fully discharged, the TFTs across the first row are re-opened. This is achieved by removing the voltage across the gate line for the first row and applying it to the gate line for the second row. The TFTs in the second row become closed and the electrical signal from each pixel (photodiode/capacitor) in the second row since the last readout cycle is then migrated to the adjacent data line (Figure 2.14) for each pixel (photodiode/capacitor).

This process is then repeated for the whole array with the TFTs in each row being closed and then opened sequentially. The electrical signal for each pixel is then amplified and fed to an ADC, which converts the electrical signals collected into a digital value of binary form.

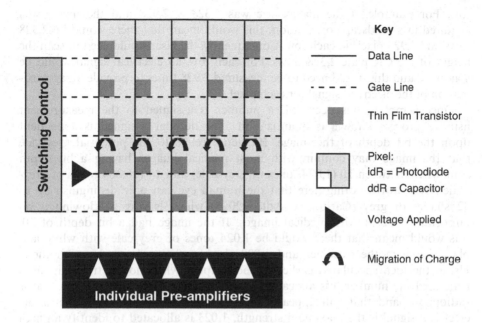

FIGURE 2.14 TFT Array: Second Row Readout.

Analogue to Digital Conversion

As speculated earlier in this chapter, if the X-ray photons reaching the image receptor have interacted with an area that is radiolucent such as air; then the electrical signal is significantly large. Conversely, if the X-ray photons reaching the image receptor have interacted with an area that is radiopaque such as bone; then the electrical signal is significantly small. Consequently, the electrical signal generated is dependent upon the density of tissue that the X-ray photons have interacted with and the intensity of X-ray photons exiting the object being imaged.

For tissue types to be visible on a digital image, the electrical signals collected by digital image receptors need to be measured. This is the first step of analogue to digital conversion and is known as sampling. 'Sampling' is almost a misnomer in digital Radiography (dR) as the electrical signal collected by each pixel (photodiode/capacitor) is measured individually due to the readout mechanism employed. With CR, a laser in a row-by-row fashion scans the CR plate. The number of rows that the laser scans is dictated by the matrix size of the CR plate. It is worth noting here that each individual size of CR plate e.g. 24 cm × 30 cm, will have a set matrix size e.g. 2,328 × 2,928 pixels. Therefore, the laser would scan the plate for the respective number of rows. Subsequently for each row, the electrical signal that is generated needs to be measured in regard to the number of pixels in that row. This means the electrical signal for a row is measured a set amount of times that would equate to the number of pixels in that row. For example, if the matrix size was 2,328 × 2,928 and the image was acquired in a landscape orientation, this would mean that there would be 2,328 rows and 2,928 pixels in each row. Consequently, the laser would need to scan the height of the CR plate 2,328 times. For each row, an electrical signal would be generated and this would need to be measured 2,928 times to provide an approximation of the electrical signal for each pixel.

Once sampling has occurred, a number is assigned to the measurement using a process known as quantization. The number assigned is dependent upon the bit depth of the image. Bit depth refers to the potential greyscale that the image may contain with most medical images having a bit depth somewhere between 10 (1,024 tones of greyscale) and 14 (16,384 tones of greyscale). It is worth noting here that the human eye can only distinguish up to 32 shades of grey (Carlton & Adler, 2013) which is why windowing is an important concept with medical images. If the image had a bit depth of 10, this would mean that there would be 1,024 tones of greyscale with white and black being at the extremes and 1,022 shades of grey being between these. Using the idea speculated earlier in the chapter, when no electrical signal is generated the number 0 is allocated in this instance to identify an area that is radiopaque and that will appear white on the image. Conversely, when the electrical signal is the maximum strength, 1,023 is allocated to identify an area that is radiolucent and that will appear black on the image. Electrical signals between these will then be allocated a number between 0 and 1,023 depending upon signal strength (Figure 2.15). For example, 256 = light grey; 512 = true grey; 768 = dark grey.

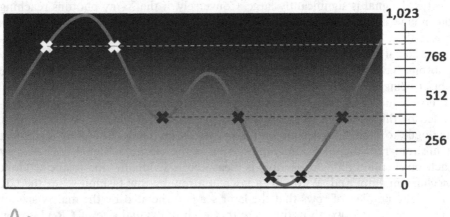

M   Electrical Signal

✕✕✕   Sampling Points

FIGURE 2.15   Quantization of an Electrical Signal.

It is important to note a number can be allocated more than once as they are purely integers that indicate the strength of the signal. Furthermore, the maximum and minimum integers here can swap in respect to signal strength and depiction of greyscale. This is seen when comparing digital imaging systems manufactured by European vendors to their American competitors. This introduces the concept of positive and negative scaling histograms in respect to image processing and enhancement.

The final part of analogue to digital conversion is to convert the numeric value assigned in the quantization stage to a binary sequence. If an image has a bit depth of 10, then the binary sequence for 0 would be 0000000000 whereas 1,023 would be 1111111111. It is this sequence of numbers that a computer uses to represent a pixel.

DIFFERENCES IN DIGITAL RADIOGRAPHY RECEPTORS

If the introduction of CR moved the acquisition of radiographic images from radiographic film to digital media, why was dR introduced?

Although CR is less costly and can be used with existing X-ray generating equipment, it is seen to have poorer spatial resolution when compared to FSR and dR. In regard to spatial resolution, there is no argument that FSR is far superior. Lança and Silva (2009b, page 135) state that spatial resolution ranges from '25–80 μm in FSR, 100–200 μm in CR and 127–200 μm in dR'. However, due to image processing facilities such as magnification, DRs are able to overcome constraints with regard to spatial resolution (Bansal, 2006).

In regard to DRs and spatial resolution, it is recognized that PSPs employed in CR tend to be inferior in terms of image quality and diagnostic value when compared with FPDs (Körner *et al.*, 2007); despite the values stated by Lança and Silva (2009b). This is primarily due to:

- The number of conversion steps which would lead to a small amount of signal being lost for each step.
- Scattering of the X-ray photon within the phosphor which leads to electron excitation occurring in areas adjacent to where the X-ray photon interacted. Subsequently, information is recorded as part of the latent image that perhaps is not a true representation of the incident intensity of X-ray photons.
- Divergence of light through the thickness of the phosphor leads to information being recorded which is not a true representation of the incident intensity of X-ray photons.
- Reading of the plate by the laser which leads to scattering of the light released which in turn produces noise and reduces spatial resolution.

The advent of needle image phosphors (NIPs) was seen as one way to reduce scattering of the X-ray photon and divergence of light through the thickness of the phosphor. NIPs have a structured phosphor where the crystals of the phosphor are shaped like needles that allow scattered photons and divergent light to be contained within the needle. However, these types of phosphors are extremely brittle and suffer from cracking due to physical weight placed upon them or excessive readout cycles.

It is also worth mentioning here that due to scattering of X-ray photons within CR, employing high kV/low mAs techniques with CR leads to noisy images with poor spatial resolution and evidence of quantum mottle. Findings by Tavares et al. (2015) demonstrate that image quality is degraded with CR, when kV is increased. This is because higher energy photons tend to scatter within the phosphor reducing spatial resolution. Reducing mAs also leads to another problem whereby having a lower intensity means there is a larger statistical difference across the PSP leading to quantum mottle. For example, as a percentage, the difference between 1 and 2 when the maximum number is 10 is 10%. However, if the maximum number increases to 100, then the difference between 1 and 2 becomes 1%. Therefore, when the plate is exposed to a lower intensity of X-ray photons, smaller differences between pixels are extrapolated. This leads to larger statistical differences which are evidenced as quantum mottle. Conversely, low kV/high mAs techniques overcome this problem, but this counteracts the principle of 'as low as reasonably practicable (ALARP)'. The use of this technique when using CR was evident in a study conducted by Vaño et al. (2007) where the average dose increased for some examinations when CR was introduced. They proposed that 'radiographers probably attempted to avoid noisy images by using milliampere-second settings higher than necessary for good image quality' (Vaño et al., 2007, page 465). A proposal that is supported by Carroll (2018) who insinuates that radiographers favor higher mAs settings to avoid quantum mottle when using CR.

One suggestion to reduce dose when employing a low kV/high mAs technique is to use additional filtration. Both Kawashima et al. (2017) and Tavares et al. (2015) propose that the use of additional filtration preferentially absorbs the lower energies within the X-ray spectrum that would not contribute to image formation. However, the use of additional filtration in CR may not be that beneficial in respect to image quality. Findings by Tavares et al. (2015) suggest that image quality is degraded when additional filtration is increased with a fixed kV selection due to a decrease in differential absorption.

In comparison to CR, the scattering of X-ray photons and divergence of light within the scintillator used in idR is minimized by employing a structured scintillator (CsI rather than GdO_2S_2). This works in a similar fashion to the structured phosphor found in NIPs. The development of nanocrystalline scintillators may also minimize these to provide high spatial resolution imaging (Luo et al., 2017). Furthermore, the introduction of irradiated side sampling (ISS) in idR has improved spatial resolution greatly. With ISS, a radiolucent TFT layer is manufactured and placed above the scintillator rather than underneath it. Consequently, when an X-ray photon passes through and strikes the scintillator, light is produced and detected immediately. When the TFT layer is below the scintillator, the light produced is then required to travel through the scintillator to reach the TFT layer. As it travels through the TFT layer, it will diverge and subsequently this reduces spatial resolution (Figure 2.16). Although structured scintillators aim to minimize the amount of divergence, a small amount of divergence will still be present with these types of scintillators.

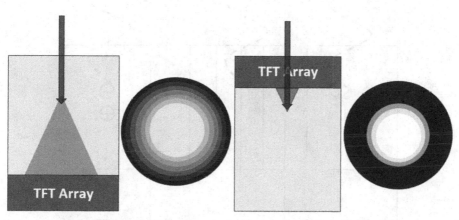

Conventional Method	Irradiated Side Sampling
Light is attenuated and diffused leading to a variance of signal onto the TFT Array which ultimately reduces spatial resolution.	Light is detected nearer to the TFT array limiting attenuation and diffusion of light. The signal is registered where it is strongest and sharpest improving spatial resolution.

FIGURE 2.16 Irradiated Side Sampling.

The issue with regard to the divergence of light is not present in ddR as the X-ray photon, upon interaction with the layer of amorphous selenium, is immediately converted into charge particles (electrical signal). Furthermore, the minimal lateral diffusion of these particles upon readout means that ddR, out of all of the DR receptors available, has the greatest spatial resolution (Figure 2.17).

However, scattering of X-ray photons within CR and to a lesser extent idR, combined with the divergence of light in both receptors, does mean that these receptors do not suffer from a phenomenon known as 'aliasing' when imaging objects of high subject contrast. Scattering of the X-ray photon and divergence of light means that noise is produced surrounding the area in which the X-ray photon interacts. As this does not happen in ddR, these receptors can suffer from severe aliasing artifacts (Kim *et al.*, 2018) when imaging objects of high subject contrast (e.g., the thorax).

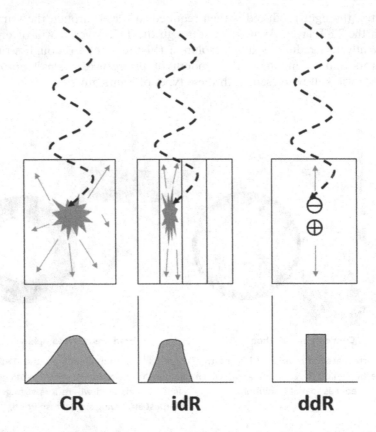

FIGURE 2.17 Comparison of CR, idR, ddR in regard to X-Ray photon absorption and its influence upon spatial resolution.

WHICH RECEPTOR IS PREFERABLE IN THE CLINICAL REALM?

Unfortunately, there is no definitive answer to this question as there are a myriad of factors to consider. These range from the type of examinations it is to be used for to financial restrictions. However, one important consideration to make in regard to the principles behind image acquisition for each receptor is detection efficiency.

Detection efficiency describes how the image receptor uses an X-ray photon of a particular energy (i.e., how much information will be obtained for the energy of the photon that interacts with the image receptor). For example, a photon of 70 keV in energy should produce information that correlates to 70 keV. In order to do this the photon would have to be completely absorbed. Unfortunately, this will not always be the case in CR as the photon energies leaving the patient are often too high in regard to the thickness of the image receptor and some will subsequently penetrate it.

Detection efficiency can be improved by

- Increasing material density – this is not seen as a viable option as it would result in an increase in the K-edge absorption value of the material. Subsequently, photons with a higher energy would be required to liberate electrons from the K-shell. This may have an impact upon spatial resolution.
- Increasing absorber thickness – this is possible but this would have a detrimental effect upon spatial resolution as it increases the distance that scatter or light within the PSP/scintillator travels.
- Using materials within the phosphor that have a slightly lower K-edge absorption value to the mean energy of an X-ray beam – although possible, this can decrease the lifetime of the receptor.

When considering the efficiency of the phosphor used in CR (BaFl:Eu^{2+} or CsBr:Eu^{2+}) and the scintillators used in idR (CsI and GdO$_2$S$_2$), they are all more efficient below 60 keV (Williams *et al.*, 2007). Considering the mean energy of an X-ray beam is 30 keV if 70 kV is selected (Carlton & Adler, 2013, page 195), it is important to consider the K-edge absorption value in respect to the mean energy of the beam as a result of the kV range used in diagnostic imaging. The K-edge absorption value of an element relates to the energy that causes a vacancy in the K-shell and usually relates to the binding energy of the K-shell for that element.

For CR, which relies upon the ionization of europium to liberate electrons that are subsequently trapped, it is worth noting that the amount of energy required to remove an electron from the K-shell of europium is 48.5 keV. However, electrons can be excited from the other shells at a considerably lower keV (6.97 keV to 8.05 keV for the L-shell; 1.13 keV to 1.80 keV for the M-shell). Similarly, for idR the K-edge absorption values for caesium, iodine and gadolinium are 35 keV, 33 keV and 50 keV respectively. Therefore, if 70 kV or more is selected then the mean energy of the beam will closely assimilate the K-edge absorption values for these receptors. The significance of this is that K-edge absorption value relates to electron excitation in CR and the release of light in scintillators for idR.

For ddR, the K-edge absorption value of selenium is approximately 13 keV. This considerable drop in K-edge absorption value means that the lifetime of this receptor, if used for general imaging, is significantly reduced. Findings by Hayre (2016) and Hayre *et al.* (2017) demonstrate that radiographers perhaps do not consider the inherent characteristics of ddR. This is evidenced in the study conducted by Hayre (2016) where some of the participants commented that they would increase kV to conform with ALARP when using this type of receptor. Although it can be surmised that this change in exposure parameters would generate a signal that is of significance in comparison to CR and idR, there is only a finite source of electrons. Once all electrons have been used, the receptor can no longer function and this is described as receptor burnout.

Consequently, it can be argued that CR and idR are more suited to the energies of photons leaving the patient in general imaging. Despite ddR having superior resolution, this particular digital imaging receptor is not particularly well suited to general radiography. However in light of the K-edge absorption value of selenium and its ability in regard to spatial resolution, ddR is far more suited to mammography where a lower kV selection (20kV–45kV) is employed and the ability to detect breast microcalcifications (<1mm) is required. It is important to state here that structures that are smaller than the detector element have to be inherently high contrast objects in order to be detected, as it is not possible to distinguish different absorption locations within a detector element.

CLINICAL IMPACT OF DR

DR has revolutionized how images are produced and the working patterns of radiology departments. Unfortunately, many of the studies in relation to the clinical impact of DR are historical with the earlier studies comparing FSR to CR and fixed dR systems. Despite this, they are still applicable in regard to modern-day working patterns, especially in respect to productivity and efficiency. Furthermore, the introduction of portable dR receptors has provided flexibility to conducting examinations (Lehnert et al., 2011).

There have been many studies that have demonstrated that both CR and dR have had a significant effect upon productivity (Andriole et al., 2002; Bacher et al., 2006; Dackiewicz et al., 2000; Körner et al., 2007; Lehnert et al., 2011; Mack et al., 2000; Reiner et al., 2005; Sack, 2001; Siegel & Reiner, 2003) and efficiency (Andriole, 2002; Mack et al., 2000; Reiner et al., 2002) due to the immediate availability and distribution of images. The majority of these studies highlight differences in examination times that focus upon the processing and post processing stages of an examination. However, Guillaume et al. (2007) propose that minimal time savings are made when comparing CR to FSR due to the movement of staff to process the PSP. Conversely, it can be argued that the processing time of a PSP is quicker than a ninety-second automatic processor used for FSR.

According to Andriole et al. (2002), the immediate availability of images for reporting or review can improve the speed-of-service. Bansal (2006) comments that images are instantly available for distribution to clinical services, and this is beneficial as it enables 'the clinical and surgical wards to begin earlier treatment and/or to safely discharge the patient due to the immediate availability of the necessary images and reports according to the relevant clinical protocols' (Nitrosi et al., 2007, page 146). Transferal of images to wards and clinics means that staff can simultaneously access the same image at the same time. Images that have been produced using FSR can only be in one place at a time while previous images would have to be retrieved from a film library, which not only can be a time-consuming process but also means that films can go missing prior to or during the filing process (Siegel & Reiner, 2003).

One consequence of the introduction of DR has been the emergence of the phenomenon known as 'dose creep'. McFadden et al. (2018, page 137) describe this as 'a phenomenon whereby radiation doses have crept upwards due to imaging staff sometimes opting to use higher exposure factors which results in a higher signal to noise ratio (SNR), producing a higher quality image with less noise'. Lança and Silva (2009a) state that it is easy to unknowingly overexpose the patient to ensure images do not demonstrate quantum mottle. It is worth mentioning here that images that are overexposed on DR systems can be manipulated to appear diagnostic. This is certainly not a recent revelation as Nol *et al.* (2006) advised that radiographers should be cautious not to develop a complacent attitude and become overconfident because of having advanced technology. For dR, this also includes the convenience of immediate viewing and having the patient in position which may also be a contributory factor to 'dose creep'; as it provides radiographers with the ability to reposition in search of a perfect image. The ability to immediately view and reposition in search of perfection is certainly evident from the findings of Hayre *et al.* (2017). It is noted that some participants believed they were unnecessarily repeating images. These findings may insinuate three contentious factors with the introduction of dR:

1. Complacency in regard to evaluating images from the perspective of whether they are answering the clinical question from the referrer.
2. Complacency in regard to initial radiographic technique as the result-ant image can be repeated if not positioned correctly.
3. Complacency in regard to exposure factor selection as the resultant image can be manipulated.

The ability to manipulate an image is primarily due to PSPs and FPDs having a wide dynamic range. Dynamic range can be described as 'the range of incident dose which can be accommodated and, therefore, contributes information to the recorded image' (Cowen *et al.*, 2008, page 491). According to Lança and Silva (2009b), the wide dynamic range in digital receptors may lead to a marked reduction of repeated radiographs and consequently a reduction in radiation exposure to the patient. This is based upon the potential to manipulate an image so long as it is acceptable with regard to radiographic positioning, relevant image criteria, and the receptor has received a sufficient amount of radiation. With digital receptors, exposure indices are used as a metric to decide whether the receptor has received a sufficient amount of radiation. For each projection and body part, exposure indices are generated and should be within the range cited by the manufacturer of the DR system for the body part and projection undertaken. This is certainly advantageous if the receptor has been slightly under or overexposed but the exposure indices are within the suggested range. However, this does raise a moral question if the exposure indices are indicative of overexposure as this would indicate that a higher than intended dose of ionizing radiation has been administered. Furthermore, what if the exposure indices are indicative of overexposure and outside the suggested range? This leads to the question whether radiographers knowingly overexpose to avoid underexposure with the knowledge that the image can be manipulated. This obviously contradicts the Ionising Radiation (Medical Exposure) Regulations (Great Britain, Department of Health, 2017a) whereby radiographers should be taking responsibility for an individual exposure while minimizing the amount of radiation through the principle of ALARP as part of the Ionising Radiations Regulations (Great Britain, Department of Health, 2017b). Consequently, it is important that radiographers do not become complacent in the course of producing high-quality images, as this could lead to patients receiving unnecessary overexposure or unnecessary repeats in search of perfection, both of which lead to an increase in radiation dose.

CONCLUSION

DRs have revolutionized the radiology department by providing an alternative form of image recording. The introduction of CR and dR has moved image acquisition into the digital era, which has overcome the historical limitations of FSR (e.g., limited exposure range, high retake rate, inflexible image display and inflexibility in film management). Furthermore, benefits have been documented regarding productivity and efficiency when using DRs.

The acquisition devices employed for CR and dR are PSPs and FPDs, respectively. PSPs are used in a similar fashion to FSR and the image is recorded by the process of electron excitation, capture and release. The inherent characteristics of latent image formation and processing of the PSP does mean that CR is inferior in terms of image quality and diagnostic value when compared with FPDs used in dR. There are two types of dR receptors available that convert incoming X-ray photons to an electrical signal either indirectly or directly. Although the former is far more suited to general radiography, as the detection efficiency for this receptor matches the mean energy of an X-ray beam at diagnostic energies, the latter provides superior spatial resolution, albeit not as superior as FSR. Despite this, it is still possible to produce images with DRs that are of equal or in some instances better quality than FSR. This can be achieved through the post-processing facilities that DR offers. Furthermore, the ability to manipulate images, so long as the patient has been positioned correctly and the PSP or FPD have received sufficient exposure, proffers a function that FSR is incapable of. However, radiographers do need to be wary in regard to the phenomenon of dose creep with DRs and the inherent complexities of using these acquisition devices.

REFERENCES

Andriole K.P. (2002) Productivity and Cost Assessment of Computed Radiography, and Screen-Film for Outpatient Chest Examinations. *Journal of Digital Imaging*, Volume 15 Number 3 Pages 161–169.
Andriole K.P., Luth D.M., & Gould R.G. (2002) Workflow Assessment of Digital versus Computed Radiography and Screen-Film in the Outpatient Environment. *Journal of Digital Imaging*, Volume 15 Number Supplement 1 Pages 124–126.
Bacher K., Smeets P., Vereecken L., De Hauwere A., Duyck P., De Man R., Verstraete K., & Thierens H. (2006) Image Quality and Radiation Dose on Digital Chest Imaging: Comparison of Amorphous Silicon and Amorphous Selenium Flat-Panel Systems. *American Journal of Roentgenology*, Volume 187 Pages 630–637.
Bansal G.J. (2006) Digital Radiography: A Comparison with Modern Conventional Imaging. *Postgraduate Medical Journal*, Volume 82 Pages 425–428.

Carlton R.R., & Adler A.M. (2013) *Radiographic Imaging: Concepts and Principles, 5th Edition*. New York: Cengage Learning.

Carroll Q.B. (2018) *Radiography in the Digital Age: Physics, Exposure, Radiation Biology*. Springfield, IL: Charles C. Thomas Publisher.

Carter C., & Vealé B. (2008) *Digital Radiography and PACS*. St. Louis, MI: Mosby Inc.

Cowen A.R., Kengyelics S.M., & Davies A.G. (2008) Solid-State, Flat-Panel, Digital Radiography Detectors and Their Physical Imaging Characteristics. *Clinical Radiology*, Volume 63 Pages 487–498.

Dackiewicz D., Bergsneider C., & Piraino D. (2000) Impact of Digital Radiography on Clinical Workflow and Patient Satisfaction. *Journal of Digital Imaging*, Volume 13 Number Supplement 1 Pages 200–201.

Fujifilm Imaging Plate Manual. (n.d.) Available – www.sb.fsu.edu/~xray/Manuals/ip.pdf

Fujifilm Medical Systems CR User Guide (2004) Available – www.spectrumxray.com/sites/default/files/pdfs/FujiFilm-CR-User-Guide.pdf

Great Britain, Department of Health. (2017a) Ionising Radiation (Medical Exposure) Regulations. Available – www.legislation.gov.uk/uksi/2017/1322/pdfs/uksi_20171322_en.pdf

Great Britain, Department of Health. (2017b) Ionising Radiations Regulations. Available – www.legislation.gov.uk/uksi/2017/1075/pdfs/uksi_20171075_en.pdf

Guillaume L., Joris T., Mandry D., Kammacher L., & Claudon M. (2007) Cost-Effectiveness Evaluation of a Digital Radiography System. *Journal de Radiologie*, Volume 88 Pages 963–967.

Hayre C.M. (2016) 'Cranking Up', 'Whacking Up' and 'Bumping Up': X-Rayexposures in Contemporary Radiographic Practice. *Radiography*, Volume 22 Pages 194–198.

Hayre C.M., Eyden A., Blackman S., & Carlton K. (2017) Image Acquisition in General Radiography: The Utilisation of DDR. *Radiography*, Volume 23 Pages 147–152.

Holmes K., Elkington M., & Harris P. (2014) *Clark's Essential Physics in Imaging for Radiographers*. Boca Raton, FL: CRC Press.

Huda W. (2016) *Review of Radiologic Physics*, 4th Edition. Philadelphia, PA: Wolters Kluwer.

Kawashima H., Ichikawa K., Nagasou D., & Hattori M. (2017) X-Raydose Reduction Using Additional Copper Filtration for Abdominal Radiography: Evaluation Using Signal Difference-to-Noise Ratio. *Physica Medica*, Volume 34 Pages 65–71.

Kim D.-S., Kim E., & Shin C-W. (2018) Oversampling Digital Radiography Imaging Based on the 2x2 Moving Average Filter for Mammography Detectors. In Lo J.Y., Schmidt T.G., Chen G.-H. Eds. Poster presented at SPIE Medical Imaging 2018, Houston, TX, United States of America, 10th-15th February 2018.

Körner M., Weber C.H., Wirth S., Pfeifer K.J., Resier M.F., & Treitl M. (2007) Advances in Digital Radiography; Physical Principles and System Overview. *Radiographics*, Volume 27 Pages 675–686.

Lança L., & Silva A. (2009a) Digital Radiography Detectors – A Technical Overview: Part 1. *Radiography*, Volume 15 Number 1 Pages 58–62.

Lança L., & Silva A. (2009b) Digital Radiography Detectors – A Technical Overview: Part 2. *Radiography*, Volume 15 Number 2 Pages 134–138.

Lança L., & Silva A. (2013) *Digital Imaging Systems for Plain Radiography*. New York, NY: Springer.

Leblans P., Vandenbroucke D., & Willems P. (2011) Storage Phosphors for Medical Imaging. *Materials*, Volume 4 Pages 1034–1086.

Lehnert T., Naguib N.N., Ackermann H., Schomerus C., Jacobi V., Balzer J.O., & Vogl T.J. (2011) Novel, Portable, Cassette-Sized, and Wireless Flat-Panel Digital Radiography System: Initial Workflow Results Versus Computed Radiography. *American Journal of Roentgenology*, Volume 196 Pages 1368–1371.

Luo Z., Moch J.G., Johnson S.S., & Chen C.-C. (2017) A Review of X-Raydetection Using Nanomaterials. *Current Nanoscience*, Volume 13 Number 4 Pages 364–372.

Mack S., Holstein J., Kleber K., & Grönemeyer D.H. (2000) New Aspects of Image Distribution and Workflow in Radiology. *Journal of Digital Imaging*, Volume 13 Number Supplement 1 Pages 17–21.

McFadden S., Roding T., de Vries G., Benwell M., Bijwaard H., & Scheurleer J. (2018) Digital Imaging and Radiographic Practise in Diagnostic Radiography: An Overview of Current Knowledge and Practice in Europe. *Radiography*, Volume 24 Pages 137–141. https://www.sor.org/system/files/article/201805/082_s_mcfadden_radiography_ 24_2018_137-141.pdf

Nitrosi A., Borasi G., Nicoli F., Modigliani G., Botti A., Bertolini M., & Notan P. (2007) A Filmless Radiology Department in a Full Digital Regional Hospital: Quantitative Evaluation of the Increased Quality and Efficiency. *Journal of Digital Imaging*, Volume 20 Number 2 Pages 140–148.

Nol J., Isouard G., & Mirecki J. (2006) Digital Repeat Analysis: Setup and Operation. *Journal of Digital Imaging*, Volume 19 Number 2 Pages 159–166.

Oakley J. (2003) *Digital Imaging – A Primer for Radiographers, Radiologists, and Health Care Professions*. London: Greenwich Medical Media.

Reiner B., Siegel E., & Carrino J.A. (2002) Workflow Optimisation: Current Trends and Future Directions. *Journal of Digital Imaging*, Volume 15 Number 3 Pages 141–152.

Reiner B.I., Siegel E.L., Hooper F.J., Siddiqui K.M., Musk A., Walker L., & Chacko A. (2005) Multi-Institutional Analysis of Computed and Direct Radiography: Part I – Technologist Productivity. *Radiology*, Volume 236 Pages 413–419.

Rowlands J.A. (2002) The Physics of Computed Radiography. *Physics in Medicine and Biology*, Volume 47 Pages R123–R166.

Sack D. (2001) Increased Productivity of a Digital Imaging System: One Hospital's Experiences. *Radiology Management*, Volume 23 Number 6 Pages 14–18.

Seeram E. (2019) *Digital Radiography: Physical Principles and Quality Control*, 2nd Edition. Singapore: Springer.

Siegel E.L. & Reiner B. (2003) Work Flow Redesign: The Key to Success When Using PACS. *Journal of Digital Imaging*, Volume 16 Number 1 Pages 164–168.

Tavares A., Lança L.J.O., & Machado N. (2015) Effect of Technical Parameters on Dose and Image Quality in a Computed Radiography System. *Poster presented at European Congress of Radiology 2015*, Vienna March 4th–8th 2015.

Vaño E., Fernández J.M., Ten J.I., Prieto C., González L., Rodríguez R., & de Las Heras H. (2007) Transition from Screen-Film to Digital Radiography: Evolution of Patient Radiation Doses at Projection Radiography. *Radiology*, Volume 243 Number 2 Pages 461–466.

Williams M.B., Krupinski E.A., Strauss K.J., Breeden W.K., Rzeszotarski M.S., Applegate K., Wyatt M., Bjork S., & Seibert J.A. (2007) Digital Radiography Image Quality: Image Acquisition. *Journal of the American College Radiology*, Volume 4 Number 6 Pages 371–388.

Zhao W., Ristic G., & Rowlands J.A. (2004) X-Ray Imaging Performance of Structured Cesium Iodide Detectors. *Medical Physics*, Volume 31 Number 9 Pages 2594–2605.

3 Image Enhancement

Kevin McHugh

DIGITAL IMAGE ENHANCEMENT

The introduction of digital receptors (DRs) has moved image acquisition into the digital era overcoming the historical limitations of Film Screen Radiography (FSR) (e.g., limited exposure range). Radiographic film is described as having a range of exposure where a diagnostic image would be produced depending upon the body part being imaged. This range is known as film latitude (Figure 3.1).

FSR involves producing an image by using an X-ray tube, intensifying screen and piece of radiographic film. The final product is a range of optical densities that are visible but varied due to the intensity of light from the intensifying screen that the radiographic film is exposed to. In order for the final product to be useful, the intensifying screen has to receive a certain amount of radiation to produce the correct amount of light. Too much radiation would lead to the intensifying screen emitting too much light. Conversely, not enough radiation would lead to the intensifying screen not emitting enough light; thus, the amount of radiation and subsequent light need to be within the film latitude (Figure 3.1).

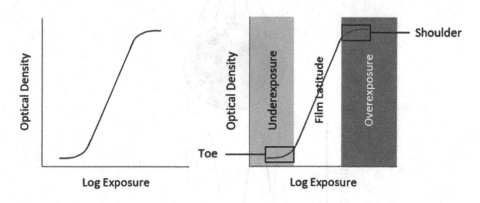

FIGURE 3.1 Characteristic Curve for FSR.

DRs are described as not having this limitation and are also described as having a wide dynamic range (latitude). This means that they are able to respond to a wide range of exposures (Figure 3.2).

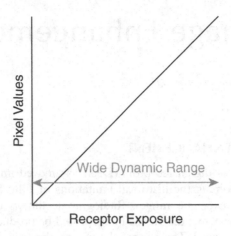

FIGURE 3.2 Digital Receptor Response to Exposure.

IMAGE HISTOGRAM

If the intensities of radiation leaving the object being imaged were to be plotted then a crude form of a histogram is produced (Figure 3.3).

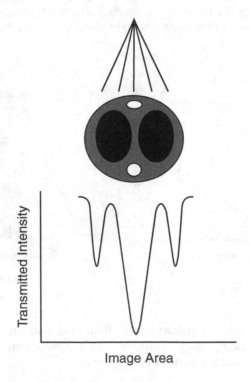

FIGURE 3.3 Intensity Histogram.

This particular histogram demonstrates the amount of exposure to the receptor in relation to the image area. For FSR, it would be imperative that this histogram is within the film latitude to avoid under or overexposure. However for DRs, it could be argued that due to the wide dynamic range, the concept of ensuring the correct amount of exposure is received by the receptor is negated. With radiographic film, under or overexposure would have been visible to the viewer. However, with DRs these can easily go unnoticed as they have the ability to correct for under and overexposure via image processing and enhancement (Seibert & Morin, 2011). This has led to experienced radiographers suggesting that the introduction of DRs has deskilled radiographers in regard to exposure factor selection (Hayre *et al.*, 2017). This is a proposition inferred by McFadden *et al.* (2018) who suggest that the selection of technical factors may play a less critical role in providing a diagnostic image due to the capabilities of DRs. Nonetheless, this is not strictly true as the receptor still requires a sufficient amount of exposure for any information to be detected and subsequently enhanced to appear diagnostic. Furthermore, knowingly overexposing the receptor to ensure sufficient information is detected cultivates a quandary from an ethical and moral perspective. Overexposure in this instance may lead to higher than intended doses and would contravene the Ionising Radiations Regulations (Great Britain, Department of Health, 2017b). Additionally, it also contradicts the Ionising Radiation (Medical Exposure) Regulations (Great Britain, Department of Health, 2017a) whereby radiographers should be taking responsibility for an individual exposure. In light of the fact that the detector still requires a sufficient amount of exposure and the role of the radiographer in regard to radiation protection legislation, it seems injudicious to claim that the introduction of DRs has led to deskilling. This ill-conceived idea possibly stems from two assumed notions:

1. Perceived automation with DRs (Hayre *et al.*, 2017) – as inferred by Seibert & Morin (2011), DRs have the ability to correct for under and overexposure. Therefore, is it possible that experienced radiographers may perceive that graduates no longer need to understand radiation protection legislation or radiation physics topics despite these still being taught and assessed at undergraduate level? This impression appears to be supported by Campbell *et al.* (2018). The findings of this study suggest that there is a negative shift in skills as new graduates or graduates trained in digital imaging only, were careless, not as knowledgeable and not performing to desired standards. Furthermore, radiographers who had been trained using FSR in this study raised concerns that 'students have lost the art of radiography' (page 5). Conversely, findings by Hayre *et al.* (2017) suggest that there is perhaps an issue with the current workforce in regard to understanding the physical principles of DRs. This may be more apparent with experienced radiographers as it is possible that the curricula that they studied did not cover this to the same extent as today. Lewis

et al. (2019) highlight that radiographers need to be educated regarding the physics and theory of DRs as inherited knowledge in regard to FSR cannot be applied when using DRs. This may potentially result in a situation where students with knowledge of DRs are working with staff who are not applying theory that students have been taught; subsequently students start to question the value of this theory to practice. Felstead & Springett (2016) propose that undergraduate students undertaking clinical placements believe that they need to adapt to the environment to learn. Therefore, this may mean that they emulate the working practices of staff which may positively or negatively influence their development. This is supported by Baldwin *et al.* (2014) who suggest it is possible for students to adopt the behaviors of individuals that they work with. With this in mind, staff are almost role models to students and 'should benchmark practice' (Felstead & Springett, 2016, page 69). Subsequently, it is the professional responsibility of radiographers who do not have the necessary knowledge, understanding, or skills regarding these technologies to ensure that they acquire them; especially as DRs become more prevalent. This is to not only establish competency with these technologies, but to ensure that radiographers are taking advantage of the benefits that DRs provide.

2. Removal of FSR acquisition and photochemistry from undergraduate curricula – these particular items have been removed and replaced with content that focuses upon the physical principles of DRs and image enhancement. This ensures that graduates have the necessary knowledge, understanding and skills that are applicable for modern-day practice as FSR becomes displaced. Consequently, the introduction of DRs has introduced new skills that radiographers need to be aware of especially in regard to digital image acquisition, digital image processing and digital image enhancement.

Image Histogram: Greyscale Histogram

For digital acquisition techniques, the intensity histogram (Figure 3.3) is shown slightly differently to demonstrate the pixel data that has been collated. The intensity or brightness of the pixels is depicted in the form of a greyscale histogram, which maps the number of pixels at each grey level present within the image. This histogram is constructed by considering the frequency of the numbers assigned in the quantization step of analogue to digital conversion (Figure 3.4).

Subsequently, it provides a convenient representation of a digital image by indicating the relative population of pixels at each brightness level and the overall intensity distribution of the image. However, these can be altered by image processing algorithms to produce corresponding changes in the image.

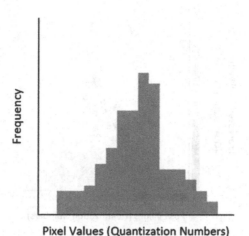

FIGURE 3.4   Greyscale Histogram.

IMAGE PROCESSING

Due to the wide dynamic range of DRs and the linear relationship between pixel values and exposure, it is necessary to apply processing algorithms to the pixel data represented in the greyscale histogram. This is because it is difficult to establish a visual correlation between image appearance and radiographic exposure (Lança & Silva, 2013). Furthermore, to process all of this data would produce an image of low radiographic contrast (Graham *et al.*, 2016; Seeram, 2019) due to the wide exposure latitude of the receptor (Seeram, 2019). Prior to image display, the pixel data is analyzed and a histogram analysis is conducted to determine the values of interest (Carlton & Adler, 2013). The values of interest are then compared with reference values for the body part and projection conducted (Carlton & Adler, 2013; Seeram, 2019). If the exposure is outside the reference values, automatic rescaling occurs in an effort to display the pixel data (Carter & Vealé, 2008). In essence, the digital imaging system locates the useful pixel data to display no matter where it is located across the latitude of the receptor.

When the operator selects the relevant body part and projection for the examination to be conducted, a processing algorithm in the form of a Look-Up Table (LUT) is applied to the pixel data. A LUT is a pre-set algorithm shaped as a characteristic curve to provide consonance to radiographic film (Figure 3.5). Consequently, the pixel data within the straight line portion of the curve will be displayed as some form of shade of grey whereas the pixel values beyond the toe and shoulder of the curve will be displayed as black and white, respectively. The y-axis of the histogram also changes to quantization number and is relabeled processed pixel values to indicate how the pixels values will be displayed.

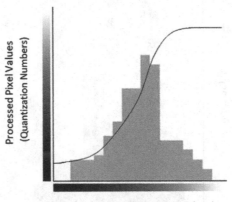

FIGURE 3.5   Greyscale Histogram: Application of LUT.

Figure 3.5 demonstrates what is known as a positive scaling histogram. In this instance the quantization number zero represents black and the largest quantization number to be allocated (dependent upon the bit depth) represents white. This is converse to what was conveyed in the previous chapter where it was speculated that if the X-ray photons reaching the image receptor have interacted with an area that is radiolucent such as air; then the electrical signal is significantly large. Consequently, the maximum quantization number is allocated to identify an area that is radiolucent and that will appear black on the image. Conversely, if the X-ray photons reaching the image receptor have interacted with an area that is radiopaque such as bone; then the electrical signal is significantly small. Consequently, when no electrical signal is generated, zero is allocated in this instance to identify an area that is radiopaque and that will appear white on the image. In clinical practice and dependent upon the manufacturer or vendor, the ends of the greyscale in respect to quantization number can be switched, introducing the concept of negative scaling histograms. The example proffered in the previous chapter was made to ease with the explanation for analogue to digital conversion and would relate to a negative scaling histogram. Logically, it makes sense for the largest quantization number to be allocated to the largest electrical signal strength and vice versa. However, it is worth remembering here that the quantization number is an integer that indicates the strength of the electrical signal during analogue to digital conversion. Subsequently, it is only being applied to indicate the degree of radiolucency or radiopacity in respect to the strength of the electrical signal. In such instances when a negative scaling histogram is present, it is found that the curve of the LUT is inverted (Figure 3.6).

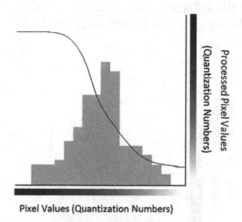

Pixel Values (Quantization Numbers)

FIGURE 3.6 Greyscale Histogram: Application of LUT (Negative Scaling Histogram).

With regard to practice in the UK, it is positive scaling histograms which are employed in regard to image enhancement. It is worth noting here the use of the term 'image enhancement' rather than 'image manipulation'. Historically, the term 'image manipulation' would have been used. However, it is possible to manipulate the image and have a detrimental effect upon its visual appearance. The term image enhancement is seen to be more fitting as 'the primary intention of the radiographer is to provide the best possible image for interpretation' (Lança & Silva, 2013, page 110).

The purpose of applying a LUT to the pixel data is to substitute new values for each pixel during image processing. For example, Figure 3.7 demonstrates how the values in a simple 3×3 matrix can be altered. Image A in Figure 3.7 represents an image of low radiographic contrast as the numerical difference between the pixel values is relatively small. However if the processing unit is instructed to replace these values then it is possible to make the difference in pixel values larger to represent an image that has high radiographic contrast (Figure 3.7: Image B).

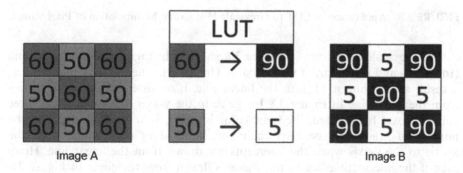

Image A   Image B

FIGURE 3.7 Application of LUT: Manipulation of Pixel Values.

To fully explore the influence that a LUT may have upon pixel data, it is necessary to consider the greyscale histogram (Figure 3.8).

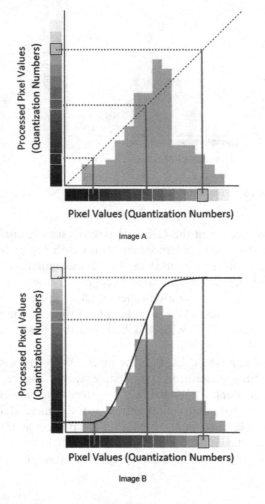

FIGURE 3.8 Application of LUT to Greyscale Histogram: Manipulation of Pixel Values.

The greyscale histogram in Figure 3.8 shows a histogram with a linear line (Image A) and a histogram with a curve (Image B). The curve is produced as a result of applying a LUT to the linear line. If an intersecting line is drawn from the linear line in Figure 3.8 Image A to the y-axis from any of the three colored boxes highlighted, the same color is noted. It is worth noting that the number of gradations between each box (four) does not change from the x-axis to the y-axis when the intercepts are drawn from the linear line. However, if the intersecting line to the y-axis is drawn from the curve in Figure 3.8 Image B, the color for each box changes. It is worth noting here that the number of gradations between each box is no longer four. To put it simply, this is how a LUT works.

By using a LUT, the data can be shaped into what the viewer deems diagnostic. Consequently, image processing alters the pixel data to produce a viewable image. Historically, radiographic images were viewed in respect to contrast and density. In FSR these were controlled by kVp and mAs, respectively. However, this does not apply to Digital Radiography, although kVp and mAs do still control radiation that exits the object being imaged. For example, the mAs still controls the number of photons that strike the detector but it has no control on the output signal as this is amplified with DRs and then subsequently altered by image processing. The kVp still has an effect on contrast and still controls the incident X-ray spectrum and therefore subject contrast and scatter. However it does not have the same influence as a LUT as the slope of the curve at every point represents how contrast will be changed. For example, if the curve has a steep slope then the image will have high radiographic contrast (large difference between pixel values).

At this point it is pertinent to highlight that 'density' is no longer an applicable term to use when viewing digital images. Density with FSR relates to optical density which is the ability of atoms to maintain energy from absorbed electromagnetic radiation. With FSR this is referred to the build-up of areas of silver due to the dissociation of silver halide molecules by X-ray photons. Consequently, density was a result of a direct physical change in the recording medium. This no longer applies with DRs as the output signal generated is not only amplified but also changed via image processing. Carlton & Adler (2013) suggest that the term density should be replaced with image receptor exposure. However, what is being viewed does not reflect the exposure that the receptor has received due to amplification of the electrical signal prior to analogue to digital conversion and application of a LUT to the pixel data. A more appropriate term to use would be 'brightness'. Although it can be argued brightness relates to a monitor control function (Carlton & Adler, 2013), it can be defined as the intensity or degree of lightness/darkness that represent individual pixels in the image (Holmes et al., 2014; Lança & Silva, 2013).

Due to the application of a LUT to the pixel data, visible feedback in regard to under or overexposure is not apparent with DRs. In order to determine whether the optimum exposure has been used, exposure indices are used as a metric. They allow the operator to decide whether the receptor has received a sufficient amount of radiation and whether the image needs repeating or can be enhanced to appear diagnostic. However, it is debatable whether manufacturers are setting exposure indices correctly (Peters & Brennan, 2002; Warren-Forward et al., 2007) or radiographers are using this metric appropriately in the clinical realm. Findings by Lewis et al. (2019) demonstrated that less than half of the participants surveyed did not know the recommended value, with a further third not knowing the corrective actions to take if the value differed from what was recommended. The authors suggested that a lack of education and appropriate training may have contributed to their findings. Although over 90% of the participants declared that they had received training in regard to digital radiography, Lewis *et al.* (2019) proposed that the training that the participants had received was not translating into acceptable application in the clinical domain.

EXPOSURE INDICES

DR systems provide an exposure index that the operator can use to assess whether the detector has been sufficiently exposed with regard to the body part being imaged. Unfortunately exposure indices are not consistent across manufacturers and vendors. They vary greatly with different names, abbreviations, mathematical formulae, calibration conditions and quantitative values being used (Mothiram *et al.*, 2014). This can cause confusion to operators when working with different systems constructed by competing manufacturers. Subsequently, 'it can be difficult to remember all the exposure values that denote a correct exposure range' (Don *et al.*, 2012, page 1337). It is beyond the scope of this chapter to consider all of the variations that are currently in practice and the reader is referred to Carlton & Adler (2013), Carter & Vealé (2008), Lança & Silva (2013), Seeram (2019) and Seibert & Morin (2011) for proprietary specific information.

The implementation of the international standard IEC 62494-1 has initiated a concerted move to standardize exposure indices between manufacturers and vendors. This standard was developed concurrently by the International Electrotechnical Commission and the American Association of Physicists in Medicine. Despite its implementation, the proprietary terms and calculations still exist. However, this does not mean that manufacturers and vendors are not adopting the international standard. In fact, they are including it alongside their own proprietary term and values.

The standard also recommended that all manufacturers and vendors should utilize a linearly proportional relationship to exposure with the aim of implementing common exposure indices (values) across all digital imaging systems. This is still not wholly evident in clinical practice primarily due to the age of existing equipment. Subsequently, it is still possible to work with equipment whereby logarithmic (e.g., Agfa) or even inversely proportional (e.g., Philips) relationships to exposure are present. Consequently, the values generated as exposure indices on older existing equipment will have a different meaning and context between systems. Therefore, it is important that radiographers are aware of these relationships to truly understand the significance of the exposure indices being generated. To aid with this the standard also suggested two new terms, target exposure index and deviation index, alongside the exposure index:

1. Exposure Index (EI) – This value is a measure of exposure (radiation) in the relevant image region (Don *et al.*, 2012; Mothiram *et al.*, 2014). The region to be measured is usually manufacturer dependent

and can vary between systems. This is usually calculated at the histogram analysis and at the same time as determining values of interest (Carlton & Adler, 2013).

2. Target Exposure Index (EI_T) – This is a reference value and acts as the optimum exposure to achieve a diagnostic image. This will be dependent upon the body part, projection, exposure parameters selected and the digital receptor used. It can be contended that the introduction of the EI_T is intended to encourage radiographers to consider the influence of exposure parameters and subsequent radiation output in regard to achieving a diagnostic image.

3. Deviation Index (DI) – This value is effectively an assessment of how far the EI is from the EI_T. In essence it is an integer expressed as a numeric value which infers whether the quantity of radiation used was correct or not (i.e., whether the receptor was under or overexposed). The numeric value cited correlates to a percentage in respect to exposure which can vary slightly between manufacturers (See Figure 3.9).

Deviation Index	Exposure Status	Percentage
> +3.0	Overexposure (double or more)	+100% or more
+1.0 to +3.0	Overexposure	+26% to +100%
+0.5 to -0.5	Target Range	
-1.0 to -3.0	Underexposed	-20% to -50%
< -3.0	Underexposed (half or even less)	-50% or less

FIGURE 3.9 Clinical Application of Deviation Index.
(Adapted from Don *et al.*, 2012)

Historically, radiographers would have used the EI to gauge whether the receptor had received a sufficient quantity of radiation to achieve a diagnostic image. Manufacturers and vendors also advised radiographers to use it as a feedback mechanism to check whether the exposure parameters used were correct and whether the image could be enhanced or not. However, it is now the DI that is the metric to observe regarding feedback in respect to the exposure parameters used. Despite this, the EI can still be used to assess whether the image could be enhanced.

IMAGE ENHANCEMENT

One advantage of capturing images digitally is that the image can be optimized with regard to contrast, brightness and general image visibility. Look-Up Tables (LUTs) are used to produce a 'characteristic' curve that fits the body part that has been imaged. For instance, a LUT for a low kVp chest will produce a curve that has a steep slope to create a resultant image that has high radiographic contrast. It can be argued here that due to the subject contrast of the chest, the image should naturally have high radiographic contrast. However, what must be remembered here is that DRs have a wide dynamic range. Unless the pixel data is focused upon, the resultant image will be of low radiographic contrast as the entire range of the receptor is taken into consideration. Although the LUT focuses upon the pixel data and fits the pixel data to the correct 'characteristic' curve, this does not necessarily mean that the image produced from the data acquired will be diagnostic to the viewer. Therefore, what if only a minor change is necessary to improve the appearance of the image after the application of a LUT?

Many authors (Carlton & Adler, 2013; Holmes *et al.*, 2014; Lança & Silva, 2013; Oakley, 2003; Seeram, 2019) highlight several optimization techniques that can be defined as point processing operations, local processing operations and global processing operations. It is beyond the scope of this chapter to discuss all of these and the reader is referred to the relevant chapters in Carlton & Adler (2013), Holmes *et al.* (2014), Lança & Silva (2013), Oakley (2003) and Seeram (2019). Nonetheless, this chapter will focus on windowing; a point processing operation that involves adjusting the pixel data to processed values (Carlton & Adler, 2013; Seeram, 2019).

Windowing

Windowing is the process of selecting some segment of the total pixel value range and then displaying the pixel values within that segment. This means that contrast will be visible only for the pixel values that are within the selected window. Carlton & Adler (2013) explain the importance of windowing as they state that 'the human visual range encompasses 32 or fewer shades of grey, whereas the photon beam exiting the patient encompasses 1,000 different variations' (page 329). The wide dynamic range of DRs means that these variations can be detected. Consequently, all pixel values that are outside of the window will be white or black. It is worth stating here that many radiographers believe that windowing can alter the contrast and density of a digital image. This is true for contrast, but what is perceived to be density is, in fact, the brightness of the image.

Windowing allows the radiographer to adjust the position and shape of the curve that has been applied to the pixel data. This is controlled by the viewer using a computer pointing device such as a mouse. Movement of the mouse laterally controls brightness while movement of the mouse vertically controls contrast.

Windowing – Brightness

Altering the position of the curve in relation to the pixel data for both positive and negative scaling histograms manipulates brightness. In Figure 3.10 the position of the curve changes from Image A through to Image C. In Images A and B, the straight line portions of the curves are toward the ends of the greyscale. Consequently, moving the curve to the right of the pixel data makes the image brighter ('right is bright') while to the left of the pixel data makes it darker. The movement of the curve and its resultant effect on brightness is the same for both positive and negative scaling histograms. The only difference between positive and negative scaling histograms is that the ends of the greyscale are switched. If the pixel data were to be processed using these two curves (Images A and B) then the resultant image would not be diagnostic as the curves are not overlying the main body of the pixel data. Image C would produce an image that has suitable brightness as the curve is superimposed over the pixel data allowing all pixel values to be displayed.

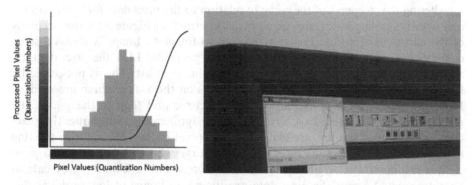

FIGURE 3.10 Application of LUT: Brightness
Image A: Curve positioned outside of the image data resulting in a bright image.

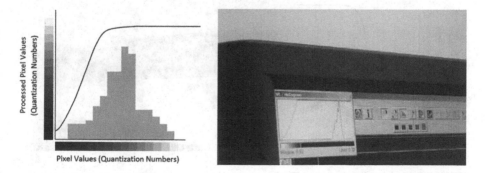

FIGURE 3.10 Image B: Curve positioned outside of the image data resulting in a dark image.

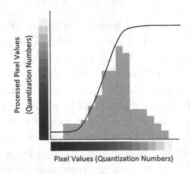

FIGURE 3.10   Image C: Curve positioned within the image data producing an image with sufficient brightness.

Windowing – Contrast

Altering the steepness of the curve in relation to the pixel data for both positive and negative scaling histograms manipulates contrast. In Figure 3.11, the steepness of the curve changes from Image A through to Image C. Image A shows a steep curve which is almost a vertical line. With this particular LUT the pixel data are altered to produce an image of high radiographic contrast. This is noticeable by looking at how the curve changes the two boxes on the x-axis which are similar. However, when intercepts are drawn to the curve and then to the y-axis, the gradation between these two boxes changes significantly. Subsequently, pixel data to the left of the line are black whereas pixel data to the right of the line are white. Image B demonstrates a shallow curve whereby the processed pixel data is similar to the original pixel data collected. This allows the greyscale to be extrapolated across the pixel data producing an image of low radiographic contrast. Image C would produce a digital image that has a good contrast range allowing all pixel values to be differentiated.

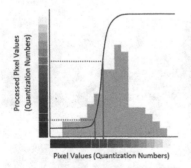

FIGURE 3.11   Application of LUT: Contrast
Image A: Steep curve positioned within the image data producing an image with high radiographic contrast.

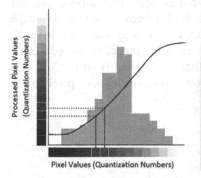

FIGURE 3.11 Image B: Shallow curve positioned within the image data producing an image with low radiographic contrast.

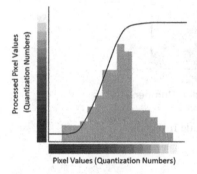

FIGURE 3.11 Image C: Curve positioned within the image data producing an image with sufficient contrast.

Windowing – Greyscale Inversion

All digital imaging systems offer the option to invert the greyscale of the image being viewed (i.e., pixels that are black become white and vice versa). Greyscale inversion is useful when viewing areas of high subject contrast or to detect objects of high radiopacity or radiolucency. This particular tool was often used in an attempt to detect minute pneumothoraces (due to the difference in brightness between the pneumothorax and the visceral line of the lung) and pulmonary nodules (De Boo et al., 2011; Lungren et al., 2012). However, recent findings regarding the use of greyscale inversion to detect pneumothoraces (Musalar et al., 2017) and pulmonary nodules (Thompson et al., 2016) refute previous evidence. Despite this, research has shown that greyscale inversion as an adjunct is useful to visualize and assess microcalcifications in mammography (Altunkeser & Körez, 2017) and provide more detailed measurements in whole spine imaging for scoliosis (Sun et al., 2016; Xia et al., 2018).

When applying greyscale inversion, the curve is inverted to the pixel data for both positive and negative scaling histograms. It does not influence the shape or the position of the curve laterally in respect to the pixel data. However, it does influence the start and end points of the curve in respect to the processed value. This in turn influences the pixel value that is assigned (Figures 3.12 and 3.13). In essence, the ends of the scale are inverted and subsequent pixel values along the scale are reassigned a new value.

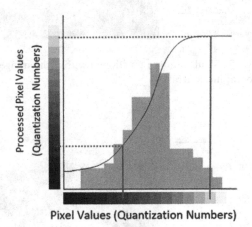

FIGURE 3.12   Application of LUT: Prior to Greyscale Inversion.

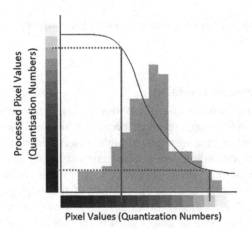

FIGURE 3.13   Application of LUT: Greyscale Inversion.

Clinical Application of Windowing

Contrast and brightness can be controlled by scrolling the mouse up and down and side to side, respectively. It is possible to visualize these changes by gaining access to the pixel data via the image processing software for the system being used. Furthermore, it is also possible at this stage to apply different LUTs to the pixel data. In some cases changing the LUT may have two consequences:

1. Changing the LUT may alter the labelling of the examination in respect to the body part and projection. This may have medicolegal implications as the image would not match the labelling of the examination and may cause confusion when the image is reported.
2. Changing the LUT may change the position and shape of the curve in regard to the pixel data (Figure 3.14).

Although the latter may not always appear to alter the visual image appearance in respect to contrast and brightness, it can (along with windowing) alter two image viewing metrics known as 'window width' and 'window level':

1. Window Width (WW) – determines the maximum number of shades of grey that are displayed on the monitor and subsequently determines contrast. A narrow WW will produce an image with high radiographic contrast but a small contrast range as there are fewer shades of grey visible (Carlton & Adler, 2013; Lança & Silva, 2013; Seeram, 2019). Conversely, a wide WW will produce an image with low radiographic contrast but a large contrast range due to an increase in the number of visible shades of grey (Carlton & Adler, 2013; Lança & Silva, 2013; Seeram, 2019).
2. Window Level (WL) – the center or midpoint of the range of pixel values can be positioned anywhere on the WW. The WL determines the brightness of the image. If the level is low then the image will appear dark whereas if it is high it appears bright (Carlton & Adler, 2013; Lança & Silva, 2013; Seeram, 2019).

WW and WL are usually pre-programmed with accordance to the LUT that is selected, but can be changed by the operator while visualizing the image either via windowing or applying a different LUT (Figure 3.14 demonstrates this as WW changes from 0.87 to 1.03 while WL changes from 2.39 to 2.47).

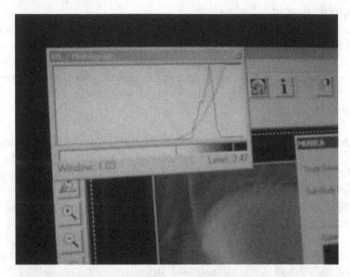

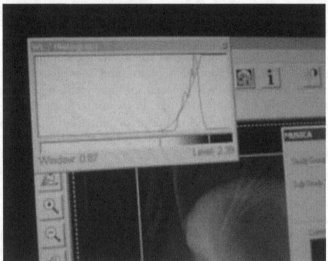

FIGURE 3.14   Changing Algorithm.

Clinical Implications of Windowing

As demonstrated in Figures 3.10 and 3.11, altering the position and shape of the curve in relation to the pixel data also alters the position of the toe and shoulder of the curve. Carlton & Adler (2013) infer that the radiographer can control which portion of the pixel data will be seen by manipulating the image. However, they forewarn 'the more the radiographer changes the original data, the less information the radiologist has to work with' (Carlton & Adler, 2013, page 330).

Earlier in this chapter it was conveyed that the pixel data within the straight line portion of the curve will be displayed as some form of shade of grey. However, the pixel values beyond the toe and shoulder of the curve will be displayed as black and white, respectively (Figure 3.15).

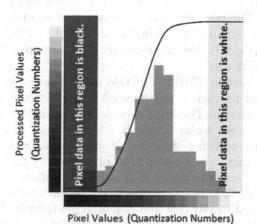

FIGURE 3.15 Application of LUT: Allocation of Black and White to Pixel Data.

If these changes were to be saved, it is possible for the pixel data beyond the toe and shoulder of the curve to be lost (erased). However this would be dependent upon whether lossy compression (some information is lost) or lossless compression (no information is lost) was employed. Subsequently, with lossy compression any pixel data beyond the toe and shoulder of the curve is deleted, as this is redundant information in regard to the dataset (Graham et al., 2005) and not part of the pixel values that are being visualized. Consequently from an image enhancement perspective, lossless compression would be ideal as the pixel data not being visualized would be kept. However, the reduction in file size when using a lossy compression technique may offer limited benefits.

The purpose of image compression is to reduce file size in respect to storage and to ease image transfer across networks. Historically, transfer and interpretation of image data between different systems was problematic as there was no commonly agreed set of rules to define the format of the data files. To overcome this problem, the Digital Imaging and Communications in Medicine (DICOM) standard was developed which defines a set of instructions when transferring image data between two different systems. DICOM groups information into data sets consisting of a header containing demographic information and the image data (Varma, 2012). Effectively DICOM can be likened to a language interpreter working with a patient who does not speak English. DICOM does not convert image data but structures the data appropriately so another system recognizes fields such as name, gender, etc. To use the analogy of a language interpreter, their role would be to ensure any communication was converted to the appropriate language but was also grammatically correct. It is the accuracy of the grammar which mirrors what DICOM effectively does as grammar is a set of structural rules within any language. Therefore, DICOM determines the structure of image data and where specific image data should be within the file (Oakley, 2003). As a result, all modalities now have DICOM compatibility to a certain extent although the level of compatibility may vary.

According to Seeram (2019), DICOM uses JPEG (Joint Photographic Experts Group) as its current compression method which offers both lossy and lossless compression. The lossy compression method works by balancing color against luminance. Color is usually down sampled to consider a smaller color palette thus reducing file size. It can be argued that using such a technique for a greyscale image is not appropriate as this could potentially reduce subjective contrast in regard to image visualization.

In light of this, there has been much discussion in the medical imaging community in regard to DICOM and potential file formats. It is important to note that DICOM is not a file format (Faccioli *et al.*, 2009; Oakley, 2003; Varma, 2012) but is an open source standard that outlines how image data will be communicated. Although Larobina & Murino (2014) state that DICOM is both a file format and network communication protocol, Varma (2012) explains that image file formats that are compliant with DICOM are described as a DICOM file format. This is further corroborated by Lança & Silva (2013) who state that a digital radiographic image is a matrix of pixels that then assembles into 'an image file hopefully compliant with DICOM' (page 110). With this in mind, Faccioli *et al.* (2009) suggest that a DICOM file can be any file format that is compliant with DICOM nomenclature (e.g., Bitmap, Tagged Image File Format and JPEG; although strictly speaking, JPEG is a compression algorithm). Consequently, the potential use of the Portable Network Graphics (PNG) file format for digital image display has been discussed. The benefits of the PNG file format are well described by Wiggins *et al.* (2001) in their seminal paper reviewing image file formats. They make it clear that the PNG file format employs a lossless compression technique which means it can be repeatedly updated and edited with the original data remaining *in situ*. In addition, its architecture allows for information to be read by certain applications and not others which would mean DICOM datasets could be embedded. Furthermore, gamma (brightness) correction allows the image to be displayed on any device or platform as an exact replica of the version on the device it was acquired from. Thus, it 'is better suited to computer generated graphics e.g. digital radiographic studies' (Wiggins *et al.*, 2001, page 796). Although there are some early implementers using PNG, it does remain to be seen whether this particular file format will be incorporated into the DICOM standard.

IMAGE ENHANCEMENT: ROLE OF THE RADIOGRAPHER

There is much debate as to whether radiographers should be windowing images prior to storage and reporting. This stems from the possibility that radiographers could change the image in a way that increases the likelihood of a false-positive or false-negative diagnosis (Larsson et al., 2007). Carlton & Adler (2013) forewarn that changes made by the radiographer may mean that radiologists have less information to work with. This is supported by Schweitzer & Berg (2010) whose findings infer that image enhancement may alter the visual information available to clinicians leading to false-positive diagnoses. However, this would only be applicable when a lossy compression technique was being used as the original pixel data would not be stored for the radiologist to revert to. If a lossless compression technique is employed, any image that is retrieved via PACS is not limited with regard to its degree of manipulation as no data has been lost. Ultimately this can enhance patient management as the image can be altered to satisfy subjective viewing conditions. However, mismanagement of patients could occur if radiographers are not encouraged to undertake image enhancement of an image employing a lossless compression technique. Geijer et al. (2007) consider image enhancement to be a natural part of reading a digital image and a necessity to view all information in the image. If radiographers do not undertake image enhancement this could lead to misinterpretation or repeating images due to a lack of perceived contrast or brightness. With regard to the latter, image enhancement would avoid an unnecessary exposure and further dose. However, the image would have to be acceptable with regard to radiographic positioning and relevant image criteria regarding diagnostic acceptability.

Krupinski (2006) suggests that image enhancement increases the time taken to interpret the image and this may not increase the percentage of correct diagnoses. However, findings by Raitz et al. (2012) suggest that the manipulation of contrast and brightness leads to a higher frequency of correct diagnoses when radiographic image criteria are applied. Consequently, image enhancement may improve the reproducibility and diagnostic value of radiographic interpretation so long as the viewer has criteria to apply to the image. Considering all radiographic images have image criteria and radiographers apply a systematic checklist for each image that they evaluate, to not conduct image enhancement (through not being permitted or neglecting to) could be seen as radiographers not fulfilling their scope of practice and eluding professional responsibility. This interpretation would be consistent with the Standards of Proficiency for Radiographers (Great Britain. Health & Care Professions Council, 2013) which infer that radiographers should:

- Understand the capability, applications, and range of technological equipment used in diagnostic imaging [Standard of Proficiency 13.13].
- Be able to appraise image information for clinical manifestations and technical accuracy, and take further action if required [Standard of Proficiency 14.10].

- Be able to manipulate exposure and image recording parameters to optimal effect [Standard of Proficiency 14.28].
- Be able to use to best effect the processing and related technology supporting imaging systems [Standard of Proficiency 14.29].
- Be able to distinguish disease and trauma processes as they manifest on diagnostic images [Standard of Proficiency 14.35].

One could argue that image enhancement is a professional responsibility as diagnostic radiographers are responsible for acquiring images to diagnose injury and disease. The Society and College of Radiographers (n.d.) extend this further by inferring that diagnostic radiographers are often the first line of image inter-pretation and this does not purely consist of a technical assessment of image quality. Diagnostic radiographers should be deciding whether images are abnor-mal or suspicious to warrant review (ibid). Subsequently, image enhancement may be required to make this decision. Therefore, it is important radiographers take ownership in regard to ensuring images are prepared correctly for report-ing. A participant in the study by Larsson *et al.* (2007) commented that employing image enhancement introduced another dimension to taking respon-sibility for the quality assurance of their own work.

There may, of course, be radiographers who are not willing or do not feel confident enough to undertake image enhancement. However, not taking the opportunity to conduct image enhancement undermines their professional status. Consequently, this does not allow diagnostic radiographers to fulfil the scope of practice that is expected of them and reinforces the perception that diagnostic radiographers are just glorified photographers. In order to avoid this label, or worse still the stigma that radiographers are merely button pushers, it is imperative that radiographers safeguard image enhancement as part of their professional responsibility.

CONCLUSION

The transition from an analogue recording medium to a digital recording medium has introduced new skills and brought many challenges to the practicing radiographer. DRs are described as having a wide latitude in comparison to Film Screen Radiography and subsequently respond to a wider exposure range. Fur-thermore, they have the inherent ability to correct under and overexposure. Con-sequently, the visible signs of under or overexposure are not as apparent as with Film Screen Radiography and radiographers need to be aware of the importance of exposure indices in regard to their practice.

DRs are able to correct under and overexposure via image processing and enhancement. Image processing involves applying a processing algorithm in the form of a Look-Up Table (LUT) to the data that has been collected by the receptor. This data is usually represented as a greyscale histogram which maps the number of pixels at each grey level (quantization number) present within the image to a frequency histogram. The LUT, which mimics the shape of a characteristic curve, is then placed over the greyscale histogram. The shape of the LUT provides consonance to radiographic film and means pixel data within the straight line portion of the curve will be displayed as some form of shade of grey. Pixel values that are beyond the toe and shoulder of the characteristic curve will be displayed as black and white, respectively. However, the shape of the characteristic curve can be altered via an image enhancement technique known as windowing. Windowing allows the radiographer to adjust the position and shape of the curve that has been applied to the pixel data. Changing the lateral position of the curve in relation to the pixel data alters brightness, while altering the steepness of the curve alters contrast. When windowing, radiographers need to be wary of any pixel data beyond the toe and shoulder of the curve, as this may be deleted if a lossy compression technique is used when saving images. Nonetheless, radiographers should not be discouraged from image enhancement to fulfil their scope of practice and professional responsibility.

REFERENCES

Altunkeser A. & Körez M.K. (2017) Usefulness of grayscale inverted images in addition to standard images in digital mammography. *BMC Medical Imaging 2017* Volume 17 Page 26. https://doi.org/10.1186/s12880-017-0196-6.

Baldwin A., Mills J., Birks M., & Budden L. (2014) Role modelling in undergraduate nursing education: An integrative literature review. *Nurse Education Today* Volume 34 Pages e18–e26.

Campbell S.S., Morton D., & Grobler A.D. (2018) Transitioning from analogue to digital imaging: Challenges of South African analogue-trained radiographers. *Radiography*, Article in press. https://doi.org/10.1016/j.radi.2018.10.001.

Carlton R.R. & Adler A.M. (2013) *Radiographic Imaging: Concepts and Principles, 5th Edition*. New York: Cengage Learning.

Carter C. & Vealé B. (2008) *Digital Radiography and PACS*. St. Louis, MI: Mosby Inc.

De Boo D.W., Uffmann M., Bipat S., Boorsma E.F., Scheerder M.J., Weber M., & Schaefer-Prokop C.M. (2011) Gray-scale reversal for the detection of pulmonary nodules on a PACS workstation. *American Journal of Roentology* Volume 197 Issue 5 Pages 1096–1100.

Don S., Whiting B.R., Rutz L.J., & Apgar B.K. (2012) New Exposure Indicators for Digital Radiography Simplified for Radiologists and Technologists. *American Journal of Roentology* Volume 199 Pages 1337–1347.

Faccioli N., Perandini S., Comai A., D'Onofrio M.D., & Mucelli R.P. (2009) Proper use of common image file formats in handling radiological images. *La Radiologia Medica* Volume 114 Issue 3 Pages 484–495.

Felstead I.S. & Springett K. (2016) An exploration of role model influence on adult nursing students' professional development: A phenomenological research study. *Nurse Education Today* Volume 37 Pages 66–70.

Geijer H., Geijer M., Forsberg L., Kheddache S., & Sund P. (2007) Comparison of colour LCD and medical-grade monochrome LCD displays in diagnostic radiology. *Journal of Digital Imaging* Volume 20 Issue 2 Pages 114–121.

Graham D.T., Cloke P., & Vosper M. (2016) *Principles and Applications of Radiological Physics, 6th Edition*. Edinburgh: Churchill Livingstone.

Graham R.N.J., Perriss R.W., & Scarsbrook A.F. (2005) DICOM demystified: A review of digital file formats and their use in radiological practice. *Clinical Radiology* Volume 60 Pages 1133–1140.

Great Britain, Department of Health. (2017a) Ionising radiation (medical exposure) regulations. Available: www.legislation.gov.uk/uksi/2017/1322/pdfs/uksi_20171322_en.pdf

Great Britain, Department of Health. (2017b) Ionising radiations regulations. Available: www.legislation.gov.uk/uksi/2017/1075/pdfs/uksi_20171075_en.pdf

Great Britain, Health & Care Professions Council (2013) Standards of Proficiency – Radiographers. Available: https://www.hcpc-uk.org/globalassets/resources/standards/standards-of-proficiency—radiographers.pdf

Hayre C.M., Eyden A., Blackman S., & Carlton K. (2017) Image acquisition in general radiography: The utilisation of DDR. *Radiography* Volume 23 Pages 147–152.

Holmes K., Elkington M., & Harris P. (2014) *Clark's Essential Physics in Imaging for Radiographers.* Boca Raton, FL: CRC Press.

Krupinski E. (2006) Technology and perception in the 21st century reading room. *Journal of the American College of Radiology* Volume 3 Issue 6 Pages 433–440.

Lança L. & Silva A. (2013) *Digital Imaging Systems for Plain Radiography.* New York, NJ: Springer.

Larobina M. & Murino L. (2014) Medical Image File Formats. *Journal of Digital Imaging* Volume 27 Pages 200–206.

Larsson W., Aspelin P., Bergquist M., Hillergård K., Jacobsson B., Lindsköld L., Wallberg J., & Lundberg N. (2007) The effects of PACS on radiographer's work practice. *Radiography* Volume 13 Pages 235–240.

Lewis S., Pieterse T., & Lawrence H. (2019) Evaluating the use of exposure indicators in digital x-ray imaging system: Gauteng South Africa. *Radiography* Article in press. https://doi.org/10.1016/j.radi.2019.01.003.

Lungren M.P., Samei E., Barnhart H., Mcadams H.P., Leder R.A., Christensen J.D., Wylie J.D., Tan J.W., Li X., & Hurwitz L.M. (2012) Gray-scale inversion radiographic display for the detection of pulmonary nodules on chest radiographs. *Clinical Imaging* Volume 36 Issue 5 Pages 515–521.

McFadden S., Roding T., de Vries G., Benwell M., Bijwaard H., & Scheurleer J. (2018) Digital imaging and radiographic practise in diagnostic radiography: An overview of current knowledge and practice in Europe. *Radiography* Volume 24 Pages 137–141.

Mothiram U., Brennan P.C., Lewis S.J., Moran B., & Robinson P. (2014) Digital radiography exposure indices: A review. *Journal of Medical Radiation Sciences* Volume 61 Pages 112–118.

Musalar E., Ekinci S., Ünek O., Ars E., Eran H.Ş., Gürses B., & Aktaş C. (2017) Conventional vs invert-grayscale x-ray for diagnosis of pneumothorax in the emergency setting. *The American Journal of Emergency Medicine* Volume 35 Issue 9 Pages 1217–1221.

Oakley J. (2003) *Digital Imaging – A primer for radiographers, radiologists, and health care professions.* London: Greenwich Medical Media.

Peters S.E. & Brennan P.C. (2002) Digital radiography: Are the manufacturers' settings too high? Optimisation of the Kodak digital radiography system with aid of the computed radiography dose index. *European Journal of Radiology* Volume 12 Issue 9 Pages 2381–2387.

Raitz R., Assunção Junior J.N.R., Fenyo-Pereira M., Correa L., & de Lima L.P. (2012) Assessment of using digital manipulation tools for diagnosing mandibular radiolucent lesions. *Dentomaxillofacial Radiology* Volume 41 Pages 203–210.

Schweitzer D.M. & Berg R.W. (2010) A digital radiographic artifact: A clinical report. *Journal of Prosthetic Dentistry* Volume 103 Issue 6 Pages 326–329.

Seeram E. (2019) *Digital Radiography: Physical Principles and Quality Control, 2nd Edition.* Singapore: Springer.

Seibert J.A. & Morin R.L. (2011) The standardized exposure index for digital radiography: An opportunity for optimization of radiation dose to the pediatric population. *Pediatric Radiology* Volume 41 Issue 5 Pages 573–581.

Society and College of Radiographers. (n.d.) *The Role of the Radiography Workforce in Image Interpretation, Film Reading and Clinical Image Reporting.* Available: www.sor.org/system/files/article/201611/image_interpretation_info.pdf

Sun W., Zhou J., Qin X., Xu L., Yuan X., Yang L., Yong Q., & Zhu Z. (2016) Grayscale inversion radiographic view provided intra and interobserver reliabilities in measuring spinopelvic parameters in asymptomatic adult population. *BMC Musculoskeletal Disorders 2016.* Available: www.bmcmusculoskeletdisord.biomedcentral.com/track/pdf/10.1186/s12891-016-1269-3

Thompson J.D., Thomas N.B., Manning D.J., & Hogg P. (2016) The impact of greyscale inversion for nodule detection in an anthropomorphic chest phantom: A free-response observer study. *British Journal of Radiology.* Available: www.birpublications.org/doi/pdf/10.1259/bjr.20160249

Varma D.R. (2012) Managing DICOM images: Tips and tricks for the radiologist. *Indian Journal of Radiology and Imaging* Volume 22 Issue 1 Pages 4–13.

Warren-Forward H., Arthur L., Hobson L., Skinner R., Watts A., Clapham K., Lou D., & Cook A. (2007) An assessment of exposure indices in computed radiography for the posterior-anterior chest and lateral lumbar spine. *The British Journal of Radiology* Volume 80 Issue 949 Pages 26–31.

Wiggins R.H., Davidson C., Harnsberger H.R., Lauman J., & Goede P.A. (2001) Image file formats: Past, present, and future. *Radiographics* Volume 21 Pages 789–798.

Xia C., Xu L., Xue B., Sheng F., Qiu Y., & Zhu Z. (2018) Grayscale Inversion View can improve the reliability for measuring proximal junctional kyphosis in adolescent idiopathic scoliosis. *World Neurosurgery* Volume 119 Pages e631–e637.

4 Dose Creep in Digital Radiography

Rob Davidson

INTRODUCTION

Exposure or dose creep is the gradual increase of radiographic exposures over time in digital radiography for the same planar radiographic examination and projection. This is a phenomenon that has only arisen since the advent and use of the DR techniques of computed radiography, indirect digital radiography, and direct digital radiography. The result of exposure or dose creep is that radiographers are increasing the dose to their individual patient above what is needed for optimal image quality. When exposure or dose creep results, the ALARA (as low as reasonably achievable/acceptable) principle has not been followed.

This chapter will examine the phenomenon of exposure or dose creep, and how it came about and the methods of monitoring dose for individual DR images and individual radiographers. Further, the chapter will examine the changes in radiographic practice are needed to address or stop exposure or dose creep occurring.

PHENOMENA OF EXPOSURE OR DOSE CREEP

FILM/SCREEN

Film/screen (F/S) radiography was in clinical use for approximately 100 years since Röntgen's discovery of X-ray in 1895. Since the early 1980s, digital radiography receptors have replaced F/S in planar radiography examinations. In F/S radiography the radiographer was required to use an appropriate focal-film distance (FFD), which is now typically called the source to image distance (SID), for distortion and tube loading issues; select an appropriate tube voltage (kVp) for image contrast; and then select the correct tube current (mAs) to provide an optimum optical density (OD) on the film. kVp selection was typically chosen for a given anatomical region and mAs was selected depending on the patient size. Incorrect combinations of kVp, mAs, and FFD lead to over or under exposure the film with high OD/dark images or low OD/light images respectively (Bushong 2001). Optimal image film OD is seen in the top row of Figure 4.1 (Veldkamp et al. 2009) at a nominal dose of 100%. Reduction and increases above are 100% are shown illustrating the effect of over and under exposure the film. Changes in the images OD are apparent as the relative dose changes.

Relative Intensity

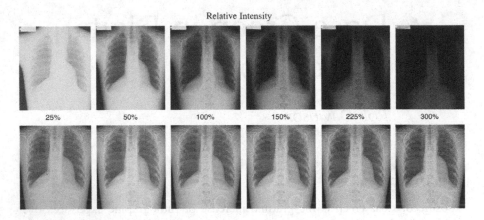

FIGURE 4.1   Digital versus film-screen chest radiography.
From (Veldkamp et al. 2009) with permission.

Differing exist intensity resulted from kVp, mAs, and FFD selection was well as the patient size. The chosen kVp, mAs, and FFD settings directly affect the dose to the patient. Figure 4.2 illustrates the film response to differing X-ray intensities as the beam exits the patient. The optimal exit intensities for the film's Hurter-Driffield (H&D) S-shaped curve response are from exposure 2 and shown between the dashed lines. This radiographic image would have a range of ODs in the image from white to black. Exit intensities from exposure 1, shown in Figure 4.2, that fall between the fine dotted lines would result in an under exposure with predominantly white ODs and be classified as a low OD image. Exit intensities that fall between the dot and dashed lines from exposure 3 would result in an over exposure, appear as a predominantly black image and be classified as a high OD image (Bushong 2001).

One of the prime goals of the radiographer using F/S systems was to optimize image OD and contrast. As such the radiation dose delivered to the patient was also optimized. When the film appearance was too white or too black incorrect, radiographers also knew that they had set incorrect radiographic factors. Achieving optimal film OD (or brightness) and contrast was feedback to the radiographer that the correct radiographic exposure factors were selected and the dose to the patient was optimized.

EXPOSURE LATITUDE OF DIGITAL RADIOGRAPHY

Computed radiography (CR) or storage phosphor systems for use in medical imaging was developed by Fuji Medical Systems and introduced into clinical practice in 1980 (Körner et al. 2007). CR was the first use of digital approaches in planar general radiographic examinations. Amorphous selenium–based image plates and amorphous silicon–caesium iodide flat-panel detectors were later

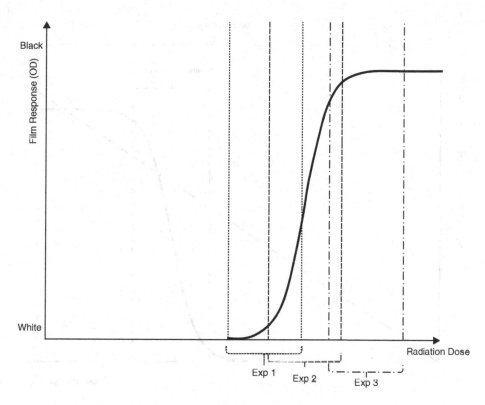

FIGURE 4.2 Film response to differing X-ray intensities, from exposures 1 to 3, as the beam exits the patient.

introduced and used clinical (Körner et al. 2007). These became known as direct digital radiography (DDR) and indirect digital radiography (IDR), respectively and collectively, IDR, DDR, and CR are referred to as digital radiography (DR).

DR systems have wider exposure latitudes or dynamic range than F/S (Seibert et al. 1996, Willis 2002, Körner et al. 2007, Seeram et al. 2013, Bushong 2016). Figure 4.3 shows typical linear responses of DR systems to radiation doses. When radiation dose or intensity is increased, the image receptor pixel values also increase. Figure 4.3 also shows the F/S response to increasing radiation dose. The F/S system must have a threshold amount of radiation before a linear response commences. As the radiation dose increases, a maximum threshold will be reached. The range of linear response by the receptor is the exposure latitude or dynamic range. The wider exposure latitudes or dynamic ranges of various DR systems means that image receptor can accept both low and high exposures and respond in a linear fashion unlike F/S that has both minimum and maximum thresholds before the images becomes white or black.

The image receptor pixels values are captured and then stored in the computer as a 'raw' file. Preprocessing operations convert the raw pixel values to

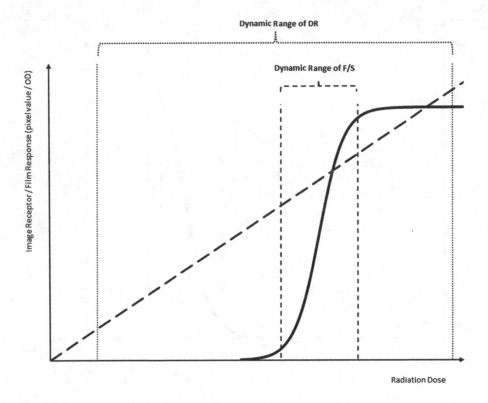

FIGURE 4.3 Dynamic range or exposure latitude of DR systems (linear dashed line) and F/S (solid S-shaped curve).

image values (Seibert 2006, Seeram 2011). During this conversion the image brightness and contrast is optimized for viewing. The correct image brightness and contrast is displayed regardless of the exposure factors set by the radiographer. The relationship between exposure factors of kVp, mAs, and SID are said to be de-coupled.

Early researchers (Freedman et al. 1993, Seibert et al. 1996, Willis 2002, 2004) noted that the large dynamic range of DR systems and the de-coupling of exposure factors and image display could lead to a potential increase of radiation dose for a given examination. It has also been noted by researchers (Seibert et al., 1996; Willis, 2004; Uffmann & Schaefer-Prokop 2009, Alsleem et al. 2014) that an increase in exposure factors, and hence dose to the patient, can increase the image quality. Signal-to-noise ratio (SNR) is the relationship between primary exit radiation (signal) and noise (scatter and other sources of noise) (Bushong 2016). An increase in SNR due to increased exposure factors increases image quality.

Conversely, a decreased of the exposure factors below a threshold can introduce noise into the image and reduce image quality. As the exit radiation's intensity from the patient reduces and/or scatter increases the SNR, and hence image quality, reduces. As seen in Figure 4.4, exposure 1 result is low pixel values which can include pixel values from noise. The SNR from exposure 1 would result in a noisy image. Noisy images may not be diagnostic. Using exposure factors for a 'standard' patient when the patient is larger than standard, reduced exit radiation, similar to exposure 1 in Figure 4.4, could result and image quality can be compromised.

To optimize radiation dose, exposure factors used should result in the exit radiation similar to exposure 2 in Figure 4.4. The exit radiation as seen from exposure 3 in Figure 4.4 will have a high SNR, it will be of high quality; however, this is will be at the cost of increased dose to the patient.

METHODS OF MONITORING RADIATION DOSE IN DIGITAL RADIOGRAPHY

Manufacturers of DR systems developed methods of measuring the amount of radiation that reached the image receptor. Collectively these methods are

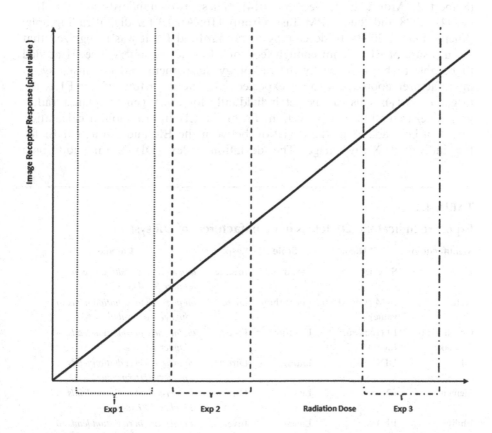

FIGURE 4.4 Linear response of DR systems with potential exit radiation does from three different exposures.

called exposure indicators (EI) (Shepard et al. 2009, Seibert & Morin 2011, Seeram, 2014). Every manufacturer of DR systems developed their version of EI. Each EI differed in its name, how it measured the amount of radiation that reached the image receptor, and the scale and type of EI units. Some examples are in Table 4.1 (Seeram 2014, Seeram et al. 2016).

Radiographers not only had to learn and understand the manufacturer's EI approach, but also had to learn what was optimal or accepted levels of exposure in the radiology department. In addition, radiographers had to understand an acceptable range of EI and what to do to modify radiographic exposures if the EI was out of the accepted range. Multiple manufacturers in a radiology department lead to multiple types of manufacturer's EI, which would have added to an increase lack of understanding. Further, Peters and Brennan (2002) noted that manufacturer EIs were set too high, and image dose could be optimized by lowering the exposure setting to obtain a lower EI value.

Two organizations, the International Electrotechnical Commission and the American Association of Physicists in Medicine, concurrently developed EI standards to overcome the issue of multiple manufacturer approaches to EIs (Seibert & Morin 2011, Seeram 2014). These two standards are the IEC 62494-1:2008 and the AAPM Task Group 116: An EI for digital radiography (Shepard et al. 2009). In developing a standardized EI, it was recognized that the measure of EI was not enough feedback for radiographers. The EI needed to be able to be modified by the radiology department and not just rely on manufacturer approaches to set expected EI. The departmental set EI is the target EI (EI_T). This can be set individually for every general planar radiographic examination and projection. From the EI_T information, a calculation can then be made as to the deviation between the EI_T and the actual EI for the individual X-ray image. The deviation index (DI) is this additional

TABLE 4.1

Exposure Indicators (EI) details of manufacturers of DR systems

Manufacturer	EI Name	Scale	Response	Comments
Fuji	S value	Linear	Inverse	*an increase in radiation leads to a lower S value*
Agfa	LgM (Log Mean value)	Logarithmic	Direct	*an increase in radiation leads to a higher LgM value*
Carestream (Kodak)	EI (Exposure Index)	Logarithmic	Direct	*an increase in radiation leads to a higher EI value*
GE	UDExp	Linear	Direct	*an increase in radiation leads to a higher UDExp value*
Siemens	EXI	Linear	Direct	*an increase in radiation leads to a higher EXI value*
Philips	EI_s	Linear	Inverse	*an increase in radiation leads to a lower EI_s value*

feedback to radiographers on their over, under, or exact meeting of the EI_T (Seibert & Morin 2011, Seeram 2014).

EXPOSURE OR DOSE CREEP

Exposure or dose creep was identified as a potential issue in digital radiography. Early researchers (Freedman et al. 1993, Seibert et al. 1996, Willis 2002) saw the dynamic range of the DR systems could allow exposure or dose creep to occur. Other researchers more formally identified exposure or dose creep (Willis 2004, Warren-Forward et al. 2007) and several authors proved the existence (Gibson & Davidson 2012). Exposure or dose creep is the gradual acceptance of higher radiographic exposures over time that then becomes the accepted or standard exposure for that X-ray examination and projection.

Gibson and Davidson (2012), in their longitudinal study, showed that exposure creep occurred where automatic exposure capture (AEC) devices were not being used. They also showed that an intervention could halt and potentially reverse the exposure creep. During their study of chest X-rays (CXR) in an intensive and critical care unit (ICCU), manual exposure factors were set. On examination of EI of over 17,000 ICCU CXR over a period of 26 months, optimal exposures decreased and over exposures increased. These trends are seen in Figure 4.5. An intervention to halt these trends was introduced. Prior to the intervention, radiographers only document the kVp and mAs of the patient's ICCU CXR in the patients' records. The intervention introduced the requirement of radiographer to document the EI of the CXR. Subsequent CXR on the same patient, radiographers had additional information on which to base their current exposure factor selection. EI of ICCU CXR was recorded for an additional 13 months following this intervention. The trend of overexposure was halted.

The first step in ensuring exposure or dose creep does not occur, or if occurring is halted and reversed, is radiographer knowledge and understanding of the effects of exposures and dose on images in DR. Hayre (2016) in his study has suggested that some radiographers lack an understanding of exposure manipulation and are relying on preset exposure values. Hayre also reported that the focus was too much on image quality at the expense of dose optimization.

Radiographers must review EI of each image as a routine part of their image and examination evaluation. Image brightness and contrast adjustments are now rarely needed due to manufacturer preprocessing algorithms. The focus of radiographers' image critique cannot just be on correct inclusion of the required anatomical region and the correct positioning of the anatomy. Image review must include dose optimization to the patient and to achieve this, the routine review of EI must be part of the image critique and review.

To ensure the ease of understanding the relationship of EI, dose and image quality, it is recommended that department management staff ensure manufacturers of the X-ray equipment used in their department display the new standardized EI and deviation index in all X-ray images.

General Radiography

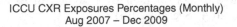

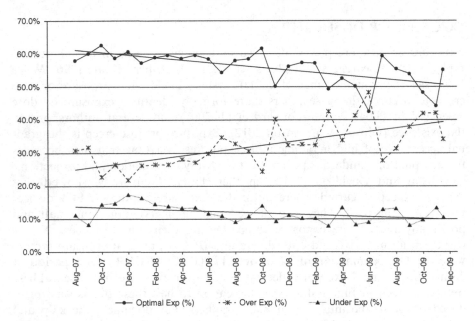

FIGURE 4.5 Over- and underexposure trend of CXR in ICCU over a 26-month period. **From (Gibson & Davidson 2012) with permission.**

AEC is used in DR examinations to terminate the exposure when a predetermined amount of radiation exit the patient and reaches the detector (Bushong 2016). The reliance by radiographers on AEC to set the appropriate amount of radiation has limitations (Doyle et al. 2005, Manning-Stanley et al. 2012). Radiographers also have control over the 'density' setting for small, average, and larger patients. When using an AEC, it is assumed that the AEC is optimized to deliver an EI with an appropriate deviation index range. As previously mentioned, researchers (Peters & Brennan 2002) have found that manufacturers' setting may not be optimized and over exposure and increased patient dose may result. Further, radiographer using an AEC approach must set the kVp. Andria et al. (2016) suggested, based on the use of test objects, that when using AEC, higher kVp setting can be used to achieve the same EI, but at less absorbed dose to the patient. Research work carried out by Zheng et al. (2016), Davidson (2016), and Jang et al. (2018) has used expert analysis of image quality through visual grading analysis and has shown that higher kVp setting can be used in DR planar X-ray examinations. Setting higher kVp or increasing the added filtration into the X-ray beam significantly lowers patient dose yet does not change the EI values of the X-ray image.

Quality assurance (QA) programs include audit of EI. David and Redden (2011) used retrospective approaches to review EI and monitor and concluded that such an approach is able to monitor exposure or dose creep. QA must go beyond programs that look at equipment or reject rates and must include review of radiographer's dose performance against departmental set standards.

Snaith (2016) in her guest editorial suggested that practice creep is a gradual change in radiographic technique occurring with or without a basis in evidence. She proposed that exposure or dose creep is an example of this. An explanation can be what Snaith described as practice drift, which is a loss of knowledge at an individual and professional level. She further proposed that this results from radiographer working more in isolation and not sharing and reviewing each other's radiographic examinations.

This phenomenon of exposure or dose creep is not just limited to planar general X-ray examinations. The issue where exposure factor increases, and hence dose, can lead to improved image quality is the same where any digital receptor is the capture device such as computed tomography (CT), digital mammography, digital fluoroscopy (DF), and digital subtraction angiography (DSA) (Honey & Hogg 2012).

CONCLUSION

Exposure or dose creep was theorized and has since been shown to exist. Radiographers are responsible for image optimization and must work under the ALARA principle. As such it is the responsibility of individual radiographers and radiology department managers to ensure everyone who exposes patient to ionizing radiation understand the phenomenon of exposure or dose creep. In gaining this understanding and knowledge, radiographer will also understand how to correct this phenomenon or to stop it from occurring. While department managers expect best practice from staff, they should be seen to be assuring best practice. In this case department managers should implement QA programs that include evaluation of EI for each X-ray examination and project. They should regularly review EI_T's and evaluate these against the ability for clinician to make a diagnosis. Importantly, patients need to be at the center of radiographers' focus on all aspects of their imaging.

REFERENCES

Alsleem, H., U. Paul, K. S. Mong and R. Davidson. (2014). "Effects of radiographic techniques on the low-contrast detail detectability performance of digital radiography systems." *Radiologic Technology* **85**(6): 614–622.

Andria, G., F. Attivissimo, G. Guglielmi, A. M. L. Lanzolla, A. Maiorana and M. Mangiantini. (2016). "Towards patient dose optimization in digital radiography." *Measurement* **79**: 331–338.

Bushong, S. (2001). *Radiologic Science for Technologists: Physics, Biology and Protection.* St Louis, MO: Mosby.

Bushong, S. (2016). *Radiologic Science for Technologists: Physics, Biology and Protection.* St Louis: Elsevier.

David, G. and A. Redden. (2011). "Retrospective analysis of computed radiography exposure reporting." *Radiologic Technology* **83**(1): 29–35.

Davidson, R. (2016). *High kVp in Abdominal X-Ray: Evaluation Using a Human Cadaver. International Society of Radiographers and Radiologic Technologists.* Seoul, Korea: ISRRT.

Doyle, P., D. Gentle and C. J. Martin. (2005). "Optimising automatic exposure control in computed radiography and the impact on patient dose." *Radiation Protection Dosimetry* **114**(1–3): 236–239.

Freedman, M. T., E. V. Pe, S. K. Mun, S.-C. B. Lo and M. C. Nelson. (1993). *Potential for Unnecessary Patient Exposure from the Use of Storage Phosphor Imaging Systems.* SPIE. SPIE Medical Imaging, 1897: 472–79.

Gibson, D. J. and R. A. Davidson. (2012). "Exposure creep in computed radiography: a longitudinal study." *Academic Radiology* **19**(4): 458–462.

Hayre, C. M. (2016). "'Cranking up', 'whacking up' and 'bumping up': x-ray exposures in contemporary radiographic practice." *Radiography* **22**(2): 194–198.

Honey, I. and P. Hogg. (2012). "Balancing radiation dose and image quality in diagnostic imaging." *Radiography* **18**(1): e1–e2.

Jang, J. S., H. J. Yang, H. J. Koo, S. H. Kim, C. R. Park, S. H. Yoon, S. Y. Shin and K.-H. Do. (2018). "Image quality assessment with dose reduction using high kVp and additional filtration for abdominal digital radiography." *Physica Medica: European Journal of Medical Physics* **50**: 46–51.

Körner, M., C. H. Weber, S. Wirth, K.-J. Pfeifer, M. F. Reiser and M. Treitl. (2007). "Advances in digital radiography: physical principles and system overview." *Radio-Graphics* **27**(3): 675–686.

Manning-Stanley, A. S., A. J. Ward and A. England. (2012). "Options for radiation dose optimisation in pelvic digital radiography: a phantom study." *Radiography* **18**(4): 256–263.

Peters, S. and P. Brennan. (2002). "Digital radiography: are the manufacturers' settings too high? optimisation of the Kodak digital radiography system with aid of the computed radiography dose index." *European Radiology* **12**(9): 2381–2387.

Seeram, E. (2011). *Digital Radiography: An Introduction.* Clifton Park, NY: Cengage Learning.

Seeram, E. (2014). "The new exposure indicator for digital radiography." *Journal of Medical Imaging and Radiation Sciences* **45**(2): 144–158.

Seeram, E., R. Davidson, S. Bushong and H. Swan. (2013). "Radiation dose optimization research: exposure technique approaches in CR imaging – a literature review." *Radiography* **19**(4): 331–338.

Seeram, E., R. Davidson, S. Bushong and H. Swan. (2016). "Optimizing the exposure indicator as a dose management strategy in computed radiography." *Radiologic Technology* **87**(4): 380–391.

Seibert, A. J. (2006). Computed radiography/digital radiography: adult. In *RSNA Categorical Course in Diagnostic Radiology Physics: From Invisibile to Visible – The Science and Practice of X-Ray Imaging and Radiation Dose Optimization.* D. P. Frush and W. Huda. Chicago, IL: Radiologic Society of North America: 57–71.

Seibert, A. J. and R. L. Morin. (2011). "The standardized exposure index for digital radiography: an opportunity for optimization of radiation dose to the pediatric population." *Pediatric radiology* **41**(5): 573–581.

Seibert, A. J., D. K. Shelton and E. H. Moore. (1996). "Computed radiography X-ray exposure trends." *Academic Radiology* **3**(4): 313–318.

Shepard, S. J., J. Wang, M. Flynn, E. Gingold, L. Goldman, K. Krugh, D. L. Leong, E. Mah, K. Ogden, D. Peck, E. Samei, J. Wang and C. E. Willis. (2009). "An exposure indicator for digital radiography: AAPM task group 116 (executive summary)." *Medical Physics* **36**(7): 2898–2914.

Snaith, B. (2016). "Evidence based radiography: is it happening or are we experiencing practice creep and practice drift?" *Radiography* **22**(4): 267–268.

Uffmann, M. and C. Schaefer-Prokop. (2009). "Digital radiography: the balance between image quality and required radiation dose." *European Journal of Radiology* **72**(2): 202–208.

Veldkamp, W. J. H., L. J. M. Kroft and J. Geleijns. (2009). "Dose and perceived image quality in chest radiography." *European Journal of Radiology* **72**(2): 209–217.

Warren-Forward, H., L. Arthur, L. Hobson, R. Skinner, A. Watts, K. Clapham, D. Lou and A. Cook. (2007). "An assessment of exposure indices in computed radiography for the posterior-anterior chest and the lateral lumbar spine." *The British Journal of Radiology* **80**(949): 26–31.

Willis, C. E. (2002). "Computed radiography: a higher dose?" *Pediatric Radiology* **32**(10): 745–750.

Willis, C. E. (2004). "Strategies for dose reduction in ordinary radiographic examinations using CR and DR." *Pediatric Radiology* **34**(3): S196–S200.

Zheng, X., M. Kim and S. Yang. (2016). *Optimal kVp in Chest Computed Radiography Using Visual Grading Scores: A Comparison between Visual Grading Characteristics and Ordinal Regression Analysis.* SPIE. SPIE Medical Imaging, 9783, MI.

5 Exposure Indicators
Black, White, or Shades of Grey

Shantel Lewis

INTRODUCTION

With the advent of digital X-ray imaging systems, besides following optimal positioning technique, the burden of optimal exposure technique selection has increased. No longer should a radiographer solely look at an image to determine if an optimal exposure technique was used. Is the image too black? Is the image too white? Now, exposure indicators, optimal collimation, use of grids, appropriate projection selection, to mention a few, are factors that radiographers must consider in order to surmise optimal exposure technique. Additionally, the hazard of x-radiation overexposure increases 500-fold, as digital X-ray imaging systems will still produce visually acceptable images at 500 times the optimal exposure (Moore *et al.*, 2012: 94).

This chapter will therefore focus on exposure indicators, beginning with the history, purpose, and factors causing variations in exposure indicators. It then discusses proprietary nomenclature of exposure indicators and efforts to standardize exposure indicators. Thereafter, the chapter delves into exposure creep and radiation protection in the digital radiography era. The chapter concludes with an analysis of the value of exposure indicators with reflection on the challenges and advantages.

HISTORY OF DIGITAL X-RAY IMAGING SYSTEMS

The production of X-rays has remained largely unchanged since its discovery yet the processing of the X-ray photons after passing through the patient has changed substantially (Behling, 2016: 5; Washington and Leaver, 2016: 109). X-ray processing and recording have progressed from wet glass plates to dry glass plates. Later flexible transparent photographic film advanced to rare earth screens combined with film which was automatically processed (Haus and Cullinan, 1989: 1203–1223; Thomas and Banerjee, 2013: 74–75). In most modern X-ray imaging departments, computer processing is currently the mainstay where images are displayed on a monitor or distributed on radiology and hospital information systems. The images may be burned to a CD, uploaded onto a USB, or printed onto laser film or paper. Images are stored digitally in a digital picture archiving and communication system (PACS) and no longer in overcrowded storage areas (Adam *et al.*, 2015: 149).

Conventional film screen radiography was largely used in the decades preceding the advent of digital X-ray imaging systems and is still in limited use today in developing countries (Bansal, 2006: 426; Thomas and Banerjee, 2013: 139–142). A general rule in conventional film screen radiography is that the product of milliampere and the exposure time (mAs) is responsible for the film blackening or density. The peak kilovoltage (kVp), also known as the quality of the X-ray beam, controls the image contrast (Curry, Dowdey and Murry, 1990: 148; Trapnell, 1967: 5). The radiographer's choice of kVp and mAs is responsible for the quantity and quality of the X-ray beam that is generated. Radiographers' knowledge of the effect of mAs and kVp on the film is therefore essential to produce optimal diagnostic films in conventional film screen radiography (Curry, Dowdey and Murry, 1990: 148; Whitley *et al.*, 2016: 41).

The resultant image density and contrast are evaluated by a radiographer to ensure optimum diagnostic quality in terms of exposure (Brake, 2016: 581; Lampignano and Kendrick, 2018: 31). If the resultant image in conventional film screen radiography appears to be too dark or too light, the radiographer retakes the image and adjusts the mAs or kVp congruently to obtain an optimum image (Long, Frank and Ehrlich, 2017: 80–81). Therefore, it becomes clear that there is a direct relationship between exposure used and the resultant image's contrast and density (Don *et al.*, 2012: 1337). In conventional film screen radiography, the film was a self-auditing tool; too dark – too much exposure was used, too light – insufficient exposure was used, optimal image – optimal exposure was used.

In the 1980s, cassette-based digital imaging X-ray systems made an appearance followed by detector-based systems in 1990 (Moore, 1980: 78; Neitzel, Maack and Günther-Kohfahl, 1994: 509). The genesis of digital X-ray imaging systems may be traced back as early as 1973 parallel with the evolution of computers, garnering use in the early 1980s and presently proliferating the developed world (Bansal, 2006: 425; Lanca and Silva, 2013: 2; Luckey, 1975). The production of X-rays remains the same in these systems, but the processing and recording of the image have changed drastically (Behling, 2016: 5; Washington and Leaver, 2016: 109).

The nomenclature for digital imaging X-ray systems differs among authors, in literature and among users (Herrmann *et al.*, 2012: 3; Schaetzing, 2003: 7; Williams *et al.*, 2007: 371). Historically, cassette-based systems have been called computed radiography (CR) or indirect digital radiography, even though technological advances see the same technology using cassette-less systems. Systems that use detectors/sensors are called digital radiography (DR) or direct digital radiography (DDR) (Campeau and Fleitz, 2017: 146; Schaetzing, 2003: 7; Williams *et al.*, 2007: 371). The focus of this chapter is not to debate the nomenclature; therefore, the historical terms are used hereafter.

COMPUTED RADIOGRAPHY

A computed radiography (CR) system consists primarily of a CR cassette, CR reader, and a computer workstation (Johnston and Fauber, 2016: 157). The mechanism of taking the X-ray for the radiographer remains unchanged (Proctor and Adams, 2014: 1060; Tighe and Brown, 2015: 16). The radiographer would still use a cassette similar to that previously used in film screen radiography. Processing of the cassette no longer occurs in the darkroom but via a CR reader. The resultant image is then displayed on a monitor for the radiographer to view, manipulate, distribute, and record (McQuille-Martensen, 2014: 8; Schaetzing, 2003: 10).

The screen is coated with a storage phosphor and a film is no longer used (Suetens, 2002: 54). The CR cassette is usually made of plastic or aluminum. The CR photostimulable phosphor plate usually consists of a protective layer, photostimulable phosphor layer, conductor layer, support layer, light shield/reflective layer, and a backing layer/base.

The protective layer cushions the photostimulable phosphor layer, which is supported by the polyester support layer to which the other layers are applied. The conductor layer prevents the build-up of static and absorbs light to increase the image sharpness. The light shield or reflective layer prevents light from erasing data or escaping from the backing. This decreases the spatial resolution. In addition, this layer reverses any light emitted from the phosphor layer in the backward direction. The base usually consists of aluminum coated with plastic to prevent scratches (Carlton and Adler, 2013: 339–340; Carrol, 2014: 596).

Certain europium-activated barium-fluorohalide compounds are used as the phosphors in CR (Carrol, 2014: 597; Sankaran, 1997: 22). When these phosphors are exposed to X-radiation, they will store the energy pattern, in F-centers, as a latent image. The CR cassette after being exposed is 'processed' by the CR reader (Gupta, Chowdhury, and Khandelwal, 2013: 207).

When the CR cassette is placed into the CR reader, the imaging plate is removed and scanned by a helium neon or semiconductor laser. Photomultipliers collect the light released from the phosphors and the signal is amplified and digitized (Bushong, 2013: 288). The digitized signal undergoes manufacturer proprietary algorithms of processing and the image is displayed on the computer workstation (Carlton and Adler, 2013: 345).

DIGITAL RADIOGRAPHY

In digital radiography (DR) a detector receives the X-ray photons and produces an image without the image storage and reading step (Torigian and Ramchandani, 2017: 4). The mechanism of taking the X-ray changes slightly for the radiographer because the use of a cassette is no longer required. Therefore, the radiographer does not need to take the cassette to be processed in a darkroom or by a CR reader. The image is displayed immediately on a monitor for viewing, manipulating, distributing, and recording (Carlton and Adler, 2013: 353).

DR, also referred to as direct digital radiography (DDR), may be a flat panel detector, two-dimensional or area charge–coupled device array, slot-scanning or photon counting type DR system (Singh, Sasane, and Lodha, 2016: 94; Verma and Indrajit, 2008: 203):

- Flat panel detector-based systems may be either direct or indirect conversion type detectors. The direct type consists of amorphous silicon photoconductors with thin film transistors which convert the X-ray photons received directly into electrical signal. The indirect type may consist of a phosphor made of either thallium doped cesium iodide, gadolinium oxysulfide, or gadox. Here the X-ray photons are first converted into light by the phosphor (Sherer *et al.*, 2018: 215; Singh, Sasane, and Lodha, 2016: 94).
- Two-dimensional or area charge coupled devices use scintillators or phosphors to convert X-ray photons into light which is directed to a charged-couple device array (Singh, Sasane, and Lodha, 2016: 94).
- Slot-scanning types use a narrow fan beam synchronized to the detector. The beam traverses the X-rayed area with two slit collimators that prevent scatter from reaching the detector (Lee *et al.*, 2009: 52).
- A photon counting type DR system is similar to the slot-scanning technique but uses a multislit detector. As the name suggests the X-ray photons that are absorbed are counted (Verma and Indrajit, 2008: 205).

The images from both CR and DR systems are displayed on monitors. Radiographers view the images on the monitor and then decide if the image is acceptable or needs to be retaken. Part of evaluating the image entails the quality of image which plays an important role when evaluating the image for acceptable diagnostic quality (Brake, 2016: 581). It is vital for radiographers working in the digital era to understand that factors affecting image quality differ between conventional and digital X-ray systems. This will allow radiographers to make informed decisions when evaluating the image for diagnostic quality.

FACTORS AFFECTING IMAGE QUALITY

Image quality in conventional film screen radiography, CR, and DR are influenced by many factors and vary according to manufacturers and X-ray systems used (Vyborny, 1997: 473). A generic discussion of some of these factors affecting image quality in conventional film screen radiography, CR, and DR is presented subsequently.

DETECTION EFFICIENCY

Detection efficiency describes how well an X-ray imaging system uses incident X-ray photons. Detection efficiency for conventional film screen radiography, CR, and DR is determined by the thickness, density, and composition of the detector that absorbs the incident X-ray photons (Williams *et al.*, 2007: 374). Denser materials with higher atomic numbers and thicker detectors will increase the efficiency (Carlton and Adler, 2013: 436).

DYNAMIC RANGE

Dynamic range (also called exposure latitude) is the ratio of the largest and smallest input X-ray intensities that can be used to produce acceptable quality. The largest intensity is determined by the saturation level of the receptor and the smallest by the intrinsic noise of the system (Körner *et al.*, 2007: 682; Samei, 2003: 45; Williams *et al.*, 2007: 375).

In conventional film screen radiography, the dynamic range depends on the processing condition, the film, emulsion, and screen characteristics (Williams *et al.*, 2007: 375). A film characteristic curve for conventional film screen radiography shows the relationship between film density and exposure and describes the contrast characteristics of film over a wide range of exposures (Bansal, 2006: 426; Scott *et al.*, 2016: 344). Areas of underexposure will fall in the toe region of the film characteristic curve where low density is observed, and areas of overexposure fall in the shoulder region and will result in a dark film. The steep straight-line portion between the toe and shoulder represents the highest level of contrast (Carlton and Adler, 2013: 303; IAEA, 2016: 19).

Juxtaposed is the linear response curve of CR were photostimulated light output is directly proportional to X-ray exposure (Artz, 1997: 14). In DR systems, the small intensity of the dynamic range is determined by the system noise and the highest intensity by the charge-holding capacity of the detector element (Körner *et al.*, 2007: 682; Williams *et al.*, 2007: 375).

SPATIAL RESOLUTION

Spatial resolution is defined as the ability to visually distinguish two adjacent structures as being separate on an image (Carrol, 2014: 225). In conventional film screen radiography spatial resolution is determined by the thickness of the screen. The thicker the screen, the greater the spatial resolution (Carter and Veale, 2014: 30). In CR, spatial resolution is determined by the scatter from the laser light in the CR reader (Rowlands, 2002: 132). In indirect conversion DR systems the spread of light during X-ray to light conversion influences spatial resolution (Körner *et al.*, 2007: 682). Narrow parallel converter columns are used to reduce the spread by increasing the absorber efficiency, thus increasing the spatial resolution. Another factor in DR is the size of the detector element; structures smaller than the detector element may not be visible unless these structures are of high contrast (Williams *et al.*, 2007: 376).

NOISE

Noise is visible by changes in an image that are not related to the area being X-rayed. Noise may be random with no distinct location or may be a fixed pattern (Huda and Abrahams, 2015: 127). In conventional film screen radiography random noise is the spatial fluctuations of the developed silver halide grains in the emulsion (Guerra, 2004: 31). In CR and DR random noise may be electronic.

In conventional film screen radiography fixed pattern noise may be spatial variation in the screen thickness. CR fixed pattern noise is related to light collection efficiency in CR plate readers. DR fixed pattern noise is related to variations among preamplifier gains (Kim and Kim, 2016: 2765). In digital imaging fixed pattern noise may be eliminated by digital post processing called flat fielding (Williams *et al.*, 2007, 377). Signal encoding errors during digitization may cause quantization noise. Quantization noise is limited by using unique analog to digital units and by introducing nonlinear amplification before digitization (Samei, 2003: 39; Williams *et al.*, 2007: 377).

Scatter radiation reduces subject contrast and decreases the signal-to-noise ratio (Fauber, 2017: 42). In CR radiography, using a grid is paramount to absorb scatter, particularly if using barium halide, which is very sensitive to scatter (Martensen, 2015: 44). When using scatter grids, exposure technique has to be modified (Carlton and Adler, 2013: 234). In DR the use of a grid will reduce both the signal and noise received by the detector. Exposure technique adjustment when using a grid in DR must preserve the signal to noise ratio (Williams *et al.*, 2007: 377). When a grid is used in conventional film screen radiography exposure technique is increased according to the grid factor to maintain the optical density (Carrol, 2014: 210). Radiographers therefore need to understand the difference in exposure technique adjustments between conventional film screen radiography and DR to ensure the ALARA principle.

RADIOGRAPHIC CONTRAST

Radiographic contrast in conventional film screen radiography is the difference in optical density between two adjacent areas on the film. In DR, it is the difference in brightness of the displayed image on a monitor. Radiographic contrast in DR may be altered independent of image acquisition (Fauber, 2017: 157; Johnston and Fauber, 2016: 108; Williams *et al.*, 2007: 378).

Subject contrast is the spectrum of X-ray intensities after attenuation by a patient (Bushong, 2013: 180). Attenuation of the X-ray beam depends on the X-ray energy spectrum, kilovoltage (kV), the total beam filtration, and the amount and type of tissue irradiated (Carlton and Adler, 2013: 409–411; Williams *et al.*, 2007: 378). The greater the kV, the less the subject contrast because the beam is attenuated more evenly over different densities and Compton scatter increases (Carrol, 2014: 302).

DETECTOR QUANTUM EFFICIENCY

Detector quantum efficiency is the efficiency of a system to maintain signal to noise ratio. Increased signal to noise ratio increases the image quality. The exposure used is inversely proportional to the detector quantum efficiency;

therefore, the higher the detector quantum efficiency, the lower the exposure for a given image quality. Additionally, detector quantum efficiency is considered by many as a standard of receptor performance encompassing detector efficiency, noise, and spatial resolution (Campeau and Fleitz, 2017: 155; Carter and Veale, 2014: 33; Williams *et al.*, 2007: 378).

MODULATION TRANSFER FUNCTION

Modulation transfer function measures object contrast at different spatial frequencies (Carlton and Adler, 2013: 428; Samei, 2003: 38). The higher the modulation transfer function, the greater the image sharpness and resolution (Samei, 2003: 38).

All of the above factors as discussed affect image quality in conventional film screen radiography, CR, and DR. However, despite this, the ability to manipulate the contrast and brightness of the image in CR and DR, allows the user to produce an acceptable image regardless of the exposure used (Sherer *et al.*, 2018: 216, 241). The disparity between analogue and digital X-ray imaging systems is confounded by its ability to disguise overexposure. High-quality X-ray images may be obtained at 500 times the exposure technique necessary to obtain an optimum quality image. (Fauber, 2017: 11; Seeram and Brennan, 2017: 52; WHO, 2012a: 5). Therefore, in the absence of visual clues of exposure an indicator of exposure was introduced.

EXPOSURE INDICATOR

The first CR systems initially used the exposure indicator (EI) '*sensitivity/ s-number*' to amend the gain to obtain the latent image (Moore, 1980: 78). Later it was used to rescale digital data for increased contrast and to compensate for exposure technique dissimilitude. Eventually, it became a mainstay to correlate exposure technique (Mothiram *et al.*, 2013: 113; Shepard *et al.*, 2009: 2899).

The EI is the numerical parameter of the relative receptor exposure or the estimated absorbed dose to the detector and is dependent on the receptor efficiency and sensitivity to incident X-rays (Costa and Pelegrino, 2014: 198; Don *et al.*, 2012: 1337; Jones *et al.*, 2015: 6664; Moore *et al.*, 2012: 94–95; Ng and Sun, 2010: 4). Seeram and Brennan (2017: 183), however, indicate that the EI is not a measure of radiation levels at the detector but rather a representation of the radiation levels at the surface of the detector achieved by converting pixel values (pixel values indicate the intensity of light photons striking the pixel).

The EI is procured from the mean detector entrance exposure, which is derived from the mean pixel value within the anatomical region of interest in the acquired image. This depends on the tube current, the total detector area irradiated, the beam attenuation, and the patient's composite attributes (Kweon *et al.*, 2012: 957–958; Uffmann and Schaefer-Prokop, 2009: 207). Exposure indicators are not direct indicators of dose but are indicators to achieve optimal images at the lowest dose possible (Carlton and Adler, 2013: 333; Seibert and Morin, 2011: 574).

The EI indicates acceptable noise levels (Mothiram *et al.*, 2013: 113). Acceptable noise levels are indicative of the image quality. Therefore, in many practices, the EI is used as a quality control tool (IAEA, 2011: 8; Takaki *et al.*, 2016: 117). Consequently, exposure indicators should be accurate, consistent, and reproducible (Shepard *et al.*, 2009: 22).

Considering the wide dynamic exposure latitude and the visual deficiency to determine exposure technique in digital X-ray imaging systems, radiographers have to forget that looking at the X-ray image alone will give information on exposure techniques used. With the EI being dubbed as the only objective indicator of optimum exposure technique (ECRI Institute, 2014: 22), radiographers need to use the EI to determine if the correct exposure technique was used to acquire the X-ray image (Brake, 2016: 581; Herrmann *et al.*, 2012: 13; Seeram *et al.*, 2013: 337). Without the EI, verification of the appropriate exposure factor selected by the radiographer is impossible (Moore *et al.*, 2012: 94–95; Sandridge, 2017: 572).

Radiographers must consider the EI obtained and any presence of quantum mottle or saturation on the X-ray image to inform exposure technique. In addition, they must consider various factors other than exposure technique which will influence the EI to make informed decisions that follow the ALARA principle (Brake, 2016: 583).

FACTORS CAUSING VARIATIONS IN EI

Prior to delving into the factors that cause variations in the EI, an explanation of how digital images are processed is necessary. Before a radiographer takes a radiograph on a digital X-ray imaging system, the anatomical area being radiographed is selected on the system. The radiographer in effect is selecting an optimal pre-stored histogram in the form of a lookup table (LUT) which will be applied to the useful signals received after exposure (Fauber, 2017: 88; Martensen, 2015: 38–39).

After exposure, in CR systems, the image area is located within the image matrix by an exposure recognition field algorithm. In DR systems, only data from exposed pixels is analyzed further. One method uses histogram analysis (Seeram, 2019: 89–90). Histogram analysis looks at the frequency of the pixel gray shade values to determine useful signals. After the predetermined LUT histogram is applied to the useful signals, automatic rescaling optimizes the image contrast. The resultant image is displayed on the monitor for the radiographer to view and perform any other necessary post-processing. The EI is determined from the midpoint of the same histogram for most manufacturers (Carrol, 2014: 547; Fauber, 2017: 88–89; Martensen, 2015: 38–39).

Since the EI is obtained from the image histogram it is prone to errors related to the histogram analysis. Therefore the factors that cause variations in the EI include (Carrol, 2014: 547; Herrmann *et al.*, 2012: 14; Jeon, 2014: 4; Martensen, 2015: 42):

- extraneous exposure information, including scatter
- exposure field recognition error

- unexpected material in field
- collimation margins not detected
- extreme underexposure or overexposure
- delay in processing
- part selection from workstation menu
- central ray centring

If the anatomical part selected from the workstation menu is incorrect, the LUT used to rescale the histogram data will be incorrect. The LUT for each anatomical part has a specific optimum grayscale and brightness level. With incorrect centring, the collimation field would have to be larger to include all the anatomy that is needed, resulting in a wider histogram (Martensen, 2015: 42).

In CR it is suggested that an optimal size CR cassette be used to eliminate extraneous exposure information. If a larger CR cassette must be used, the area of interest must be centered to the CR cassette with four-sided collimation equally distributed from the cassette borders. CR cassettes must be moved away from the radiation area and be processed as soon as possible to prevent errors in the histogram analysis of the fog-altered pixel values (Lanca and Silva, 2008: 117; Martensen, 2015: 43–44).

Mothiram *et al.* (2013: 120) also identify four factors that may cause a variation in the EI:

1. Patient gender
2. Time/date of exposure
3. Grid usage
4. Presence of implant or prosthesis

Their study found that female patients had a higher variant EI, which is congruent with Lanca and Silva's (2008: 117) findings. This finding was attributed to lack of exposure technique chart optimization in both studies.

X-rays taken outside of normal working hours also showed an increase in the detector dose variations (Mothiram *et al.*, 2013: 120) similar to findings in Peters and Brennan's (2002: 2386) study. The studies attributed the findings to the level of experience of radiographers working during shifts and the choice to rather overexpose the patient than to repeat the X-ray in staff-strapped busy environments (Mothiram *et al.*, 2013: 120; Peters and Brennan, 2002: 2386).

When grids were used the EI obtained varied from that obtained when no grid was used for the same exam (Mothiram *et al.*, 2013: 120). The presence of artifacts, implants, or prostheses will widen the histogram and display a varied EI value (Martensen, 2015: 41).

The variations in the EI are caused by errors in the histogram analysis and not because of exposure to the imaging receptor. Therefore, radiographers need to be aware of factors that may cause the EI to vary. If any of these factors are present, the detector dose no longer correlates to the exposure to the imaging receptor. Radiographers need to look for any visible quantum mottle to determine if the image would be acceptable (Fauber, 2017: 90; Martensen,

2015: 41; Herrmann *et al.*, 2012: 14). It should be noted however that radiographers may not always have access to high resolution monitors and may miss subtle quantum mottle.

Shepard *et al.* (2009: 22) stress that the accuracy of the EI depends on proper calibration to a specific set of exposure conditions and the ability to determine the region of interest under study. Apart from the accuracy of the EI, mathematical algorithms used should be '*robust enough to provide a reasonably accurate and reliable estimate of the exposure indicator regardless of collimation boundaries, anatomical positioning, inclusion of foreign bodies, etc., but this is not always the case*' (Shepard *et al.*, 2009: 22).

Another conundrum faced by radiographers when using EIs is the varied nomenclature used by manufacturers. Shepard *et al.* (2009: 22), however, do note for all manufacturers this value should reflect their system's sensitivity for a given exposure. Table 5.1 below indicates the various names manufacturers use for the EI.

Agfa uses log median (lgM), which is obtained from the median value in the histogram after high and low spikes from background radiation and collimated area are eliminated. Calibration is done at an exposure of 20 micro-Grays to

TABLE 5.1

Proprietary Nomenclature of EI (Carrol, 2014: 540; Don et al., 2012: 1338; Shepard et al., 2009: 23)

Manufacturer	Exposure Indicator Name	Symbol
Agfa	Log of median	LgM
Alara CR	Exposure indicator value	EIV
Canon	Reached exposure value	REX
Fuji	S number (speed or sensitivity)	S
General Electric (GE)	Detector exposure index	DEI
GE	Entrance skin exposure	ESE
GE	Dose area product	DAP
Hologic	Exam factor, center of mass of log E histogram	
Hologic	Dose area product	DAP
iCReo	Exposure index	
Imaging Dynamics Corporation	Accutech	f#
Imaging Dynamics Corporation	Log of median of histogram	
Carestream/Kodak	Exposure index	EI
Konica	Sensitivity number	S
Philips	Kerma air product	KAP
Philips	Exposure index	EI
Shimadzu	Reached exposure value	REX
Siemens	Exposure index	EXI
Swissray	Dose indicator	DI

the imaging receptor at 75 kilovoltage peak (kVp) and 1.5mm of added copper filtration. Log median is proportional to the exposure to the imaging receptor and any changes are noted logarithmically. For a logarithm a change of 0.3 would mean double or half the exposure to the image receptor. If the lgM changes from 2.5 to 2.8 it changes by 0.3 equaling a factor of 2. This means that the exposure to the receptor is 2 times more. The lgM is defined as:

$$lgM = 2 \times log(SAL) - 3.9478$$

where the scanning average level (SAL) is the average grayscale value represented by SAL 200 = 1,214 X (exposure in milli Roentgen [mR]) or the mean output pixel value or mean digital signal (Agfa-Gevaert, 1998: 5; Carlton and Adler, 2013: 335; Khotle et al., 2009: 251; Koen, Herbst and Rae, 2008: 28; Shepard et al., 2009: 25).

In addition, the system will record the lgM values for 50 consecutive images in each imaging category and then obtain an average. The average will be the target EI. All exposures will be compared to this average and any deviations will be shown on a bar graph (Carrol, 2014: 542; Fauber, 2017: 89; Shepard et al., 2009: 25). Another factor to consider is speed class, which in the Agfa system is selected by the radiographer. The target EI will vary with the speed class selected. However, Lanca and Silva (2008: 116) indicate that over all speed classes selected, the target EI should be around 1.96.

Fuji sensitivity (S-number) was the first developed EI and was aligned with film screen speed. The S-number is inversely proportional to the exposure to the imaging receptor and is calculated by S = 200/(exposure in mR) from the center of the usable portion of the histogram. Calibration is done at an exposure of 1mR to the imaging receptor at 80 kVp with 3 mm aluminum filtration. The S-number on the Fuji system that is displayed initially reflects the EI; thereafter the S-number becomes the brightness control. This means that the S-number will change as the brightness is increased or decreased. Therefore, only the initial S-number obtained after exposure is considered as the EI (Carlton and Adler, 2013: 334; Carrol, 2014: 543; Fauber, 2017: 89; Shepard et al., 2009: 20–25; Siegel and Kolodner, 2006: 149). Seeram et al. (2016: 385) have recommended that the inverse S-number be displayed to ease radiographer understanding. This would provide a proportional as opposed to inverse correlation between the EI and the exposure technique used.

General Electric has a detector exposure index (DEI) that is proportional to the exposure at the imaging receptor. The anatomical area is first defined and the DEI is calculated from the median pixel value. Upper and lower limit DEI and the actual DEI obtained are displayed on a bar graph. These limits may be customized by the user (Carrol, 2014: 543). Jeon (2014: 2) explains that from 2011 products released from General Electric feature a newly designed DEI aligned with the requirements for exposure index of digital imaging systems by the International Electrotechnical Commission (IEC) 62494–1:2008. A deviation index indicating the deviation of the actual EI varying from an established target is displayed together with the actual EI.

Carestream uses an EI that is obtained from the midpoint of the histogram. The EI is proportional to the exposure to the image receptor and is expressed logarithmically and calculated by the formula (Carrol, 2014: 541):

$$1000 \times \log(\text{exposure in mR}) + 2000$$

Calibration is done at an exposure of 1mR to the imaging receptor at 80 kVp and 1.5 mm of aluminum filtration and 0.5mm of added copper filtration. As this is a logarithmic scale, if the EI changes by 300, the actual exposure changes by a factor of 2. If the speed class selected is 200, then the middle of the histogram would be 2,000 which would be the ideal exposure to the image receptor. At this speed class, if the EI received is 2,300 that means that the image receptor has received twice the ideal exposure. If the EI received is 1,700, that means that the image receptor has received half of the ideal exposure (Carrol, 2014: 541; Fauber, 2017: 89; Gallet, 2010: 1–2). Pesce (2016) also mentions that the EI for Carestream is inversely proportional to the speed class used; that is, for every 300 increase in the EI, the speed class is reduced by half.

Konica uses the S-value which is inversely proportional to the exposure at the image receptor and is calibrated at 80 kV. The S-value is calculated by the formula:

$$S = QR \times R/R'$$

Where QR: range of the quantization region.

R: the incident X-ray dose on the detector that produces a signal value of 1,535 (i.e., when the QR is 200 when fixed processing produces an output density of 1.2).

R': actual X-ray dose required to produce a film output density of 1.2 at a pixel following gradation processing.

Doubling of X-ray exposure under the same conditions will halve the S-value. The S-value is generated from pixel values after gradation processing and therefore is not only determined by the amount of X-ray exposure (Konica Minolta Medical and Graphic, 2010: 4–15, 4–16; Shepard et al., 2009: 32).

For their flat panel Digital Diagnost DR system, Philips uses the term 'exposure index', which is inversely proportional to the exposure at the image receptor and is obtained from the 'characteristic' pixel value of the image. The characteristic pixel value, depending on the software, is obtained from either, the average or median pixel value or other algorithms, from the determined region of interest. The formula used to calculate EI is:

$$1000/K \text{ where } K = PVc/SENS$$

PVc: the air kerma from the characteristic pixel

SENS: sensitivity of the detector

Furthermore a kV correction factor may be applied depending on the software to offset changes to the SENS caused by changes in the X-ray photon energy.

The EI follows the ISO R'10 scale. Each factor on this scale, which mimics conventional film screen speeds, corresponds to a factor of $10^{0.1}$ (Shepard et al., 2009: 34–36).

Siemens expresses EI as exposure index or EXI. It is directly proportional to the exposure at the image receptor. The EXI is calculated by using the average pixel value in central segment of a 3 × 3 matrix of the exposed area and is calibrated at 70 kV with 0.6 mm copper added filtration (Shepard *et al.*, 2009: 42).

These distinct manufacturer names and calculations of EIs may present a dilemma for radiographers who often work in a department that has equipment from many manufacturers. So in addition to understanding the EI, radiographers also have to know the EIs of the varied equipment in their department and understand the significance. To complicate things further radiographers also need to understand the factors affecting the EI in the array of different manufacturer EIs (Brake, 2016: 584).

The IEC and American Association of Physicists in Medicine (AAPM) together have tried to alleviate some of the burden of understanding EIs by proposing standardization. It is unclear though of manufacturers' compliance in providing standardized EIs (AAPM, 2009; Seeram, 2019: 87–93).

STANDARDIZATION OF EI

Simultaneous but independent efforts by the IEC and AAPM to standardize EIs have given rise to IEC 62494–1 standard and the AAPM report 116 (ACR-AAPM-SIIM, *2014*: 6; Herrmann *et al.*, 2012: 13; IEC, 2008; Seeram, 2019: 87; Seibert and Morin, 2011: 574; Shepard *et al.*, 2009). Jones *et al.* (2015: 6664) believe that manufacturers are more likely to adopt the IEC 62494–1 standard (IEC, 2008). Moore *et al.* (2012: 96) also report that at the 2010 Image Gently summit stakeholders agreed to follow the IEC 62494–1 standard. Vastagh (2011: 566) reports that the Medical Imaging and Technology Alliance (MITA) also endorses the IEC 62494–1 standard.

Three important IEC 62494–1 standardized terms that a radiographer should know are exposure index (EI), target exposure index (EI_T), and deviation index (DI).

EI is the radiation to the receptor over the region of interest as determined by the operating system or radiographer. EI is calibrated specifically and is proportional to the radiation exposure to the receptor.

EI_T is the exposure on the imaging receptor that correlates to optimal exposure of an image. The EI_T may be set by the manufacturer of the facility.

DI is the deviation index between the EI and EI_T, and is expressed logarithmically. A DI of 0 means that optimal exposure was used. Any deviation from the target exposure will correspond with an incorrect exposure to the receptor (IEC, 2008).

Radiographers have to only look at the DI across the various manufacturers in the department to determine optimal exposure. A DI value ±1.0 corresponds to one step in milliampere second (Don *et al.*, 2012: 1338; IEC, 2008; Seeram, 2019: 91).

Incumbent to obtaining correct feedback of EIs, is the radiographers' selection of the correct body part and radiographic projection, to ensure that the correct

EI_T value is used for calculation of the DI value. The calculation of EI is still influenced, as discussed earlier, by:

- part thickness
- artifacts, that is, lead shields, metal implants, and pacemakers
- source-to-image distance
- collimation
- grids
- centring
- image plate size
- equipment design
- detector design and sensitivity

It must be stressed that repeating of the radiograph on the merit of the EI alone without considering the noise in the image will lead to increased exposure to the patient. Therefore besides understanding the new standard, radiographers, must balance the need to repeat an exposure together with evaluating the noise of the image obtained (Don *et al.*, 2012: 1339; Moore *et al.*, 2012: 96; Mothiram *et al.*, 2014: 114).

These efforts of standardization will hopefully assist radiographers in better understanding the EI. Instead of radiographers having to learn a myriad of manufacturer-dependent detector dose terminology, they would only have to learn the standardized terminology. In addition, it is hoped that because of the ease of understanding the new standardized terminology across all equipment manufacturers, radiographers would actually understand and use the EI in clinical practice. Studies have shown that radiographers have a limited understanding and use of EIs as well as a lack of optimizing exposure technique and collimation in digital X-ray imaging systems (Campbell, Morton, and Grobler, 2019: e41; Hayre *et al.*, 2019: 235; Lewis, Pieterse, and Lawrence, 2019a:2; Mc Fadden *et al.*, 2018: 140).

RADIATION PROTECTION AND EXPOSURE CREEP

Combating X-radiation overexposure has led to worldwide radiation protection measures which encapsulate justifying and optimizing medical radiation exposure to ensure the as low as reasonably achievable (ALARA) principle. Central to ensuring patient protection is radiographers. Radiographers are frontline radiation workers who can ensure justification of an X-ray request and control exposure to the patient by the selection of optimal exposure and positioning techniques (ISRRT, 2018a, 2018b).

With digital X-ray imaging systems besides using optimal positioning technique, the wide dynamic exposure latitude creates a disconnect between the exposure technique used and the resultant radiograph, further complicating radiation protection measures (Brake, 2016: 581; Fauber, 2017: 11; Seeram and Brennan, 2017: 52). Radiographers have a continued obligation to ensure that radiation protection measures are implemented. However, this continues to pose challenges (Sherer *et al.*, 2018: 3–4; Szarmach *et al.*, 2015: 61).

Studies indicate that radiographers require more training in radiation protection because their educational level was found to be proportional to attitude and awareness of radiation protection (Mojiri and Moghimbeigi, 2011: 4–5; Paolicchi *et al.*, 2016: 238; Shah *et al.*, 2007: 171; Talab *et al.*, 2016: 94). A large proportion of Togolese radiographers admitted to not asking patients their last menstrual date, not using lead protection, and irradiating pregnant women, findings congruent with those in the Ivory Coast (Adambounou *et al.*, 2015: 3; Kouamé *et al.*, 2012: 558). The Bonn Call for Action 4 calls for strengthening '*radiation protection education and training of health professionals*' (World Health Organisation (WHO), 2012a: 2). Radiographers' scientific knowledge and understanding together with education in the technical advances and capabilities in digital radiography is needed to create optimal image quality at the lowest possible exposure (Gibson and Davidson, 2012: 460; Moore *et al.*, 2012: 93, 95; Seibert and Morin, 2011). This is especially true with continued evidence of exposure creep in radiography (Seeram, 2019: 88).

Even though the wide exposure latitude of digital X-ray imaging systems limits the number of repeat X-rays done, it has also shown evidence of exposure creep. The ECRI Institute (2014, 21) listed exposure creep as number 7 on its top 10 health technology hazards for 2015. The institute explains that exposure creep is an '*unintentional consequence*' of transitioning from conventional film screen radiography to digital radiography.

Studies evaluating exposure creep in digital X-ray imaging systems include a retrospective examination of 100 posteroanterior chests and 100 lateral lumbar spines which showed 7.1% exposure creep over the study time (Warren-Forward *et al.*, 2007: 31). Furthermore, a study which evaluated EIs of 1,884 consecutive neonatal chest images did not reveal any exposure creep within the study period (Cohen *et al.*, 2011: 600). Neonatal portable images at four academic hospitals were also shown to be within half or double the target exposure range (Cohen *et al.*, 2012: 672). Retrospective analysis of 5,000 adult chests, abdomen, and pelvis EIs revealed that most examinations demonstrated EIs outside the MR ranges (Mothiram *et al.*, 2014: 119). In addition, a study which compared EI's for two different systems at two different hospitals for chest, spine, abdomen, and extremities found that most radiographs were underexposed (Ribeiro *et al.*, 2016: S349). A study showed that of 1,422 retrospectively collected EI, only half the EI indicated optimal exposure technique and from the remaining half more indicated overexposure than underexposure (Lewis, Pieterse and Lawrence, 2019b: 40).

To guide optimum exposure technique, manufacturers provide a recommended EI. However, studies have found that manufacturers have set these standards too high. Recommendations are that radiology departments individually interrogate these standards to develop a balance between exposure technique and image quality to achieve the ALARA principle and possibly halt exposure creep (Peters and Brennan, 2002: 2381–2387; Seeram, 2019: 93; Warren-Forward *et al.*, 2007: 26–31).

A recommendation that emerged during investigations into exposure creep was from a study that examined exposure indicators in the emergency department and the intensive and critical care units over a 43-month period. In the first 29 months the researchers compared the EI to the standard and found evidence of overexposure in the intensive and critical care unit. For the remaining 14 months of the study an intervention strategy was implemented. This strategy required radiographers to document the exposure techniques used and the EIs obtained for each study. Following the intervention, exposure creep halted (Gibson and Davidson, 2012: 460).

The halt in exposure creep was attributed to radiographers altering their exposure technique to obtain the optimal EI. Radiographers used the previous EI values when X-raying the same patient to alter exposures to prevent over exposure (Gibson and Davidson, 2012: 460). Subsequently, Takakia *et al.* (2019) have developed a software which '*automatically calculates mAs for subsequent X-ray examinations based on mAs and DI values from prior examinations.*' The authors found that a 39% improvement in the DI compliance with the introduction of the software for intensive care unit portable chest X-rays.

The European Commission Dimond III final report on image quality and dose management for digital radiography offers a further strategy (Busch, 2004: 7). Optimization, objectivation, and standardization are suggested as steps involved in a three-level concept of image quality. Image quality is defined as high, medium, and low (Busch, 2004: 66–67). For better understanding an example from the report will be discussed. Image quality levels for skeletal radiography may be defined as (Busch, 2004: 67):

High: Diagnosis of a fracture.
Medium: Control of the position of fractures.
Low: Locating a metal object, measurements of the spine.

These image quality levels suggest that for diagnosis of a fracture, a high-quality image would be needed. Therefore, optimal exposure should be used to obtain an optimal diagnostic image to view the fracture. Once the fracture has been confirmed and treated, the follow-up X-ray no longer requires such high image quality because it is merely to determine if the fracture was reduced adequately; therefore, this image quality falls into the medium-quality level.

The medium-quality level image would in effect require a decrease in the optimum exposure technique and hence a decrease in dose to the patient. Low-quality levels will also require even a further reduction in the exposure technique hence a further reduction in the dose to the patient. In this example low-level image quality would be for a patient that had a complete replacement of, say, the hip or knee. The report suggests clinical trials be conducted to test these strategies to develop methodical frameworks for optimization, objectivation, and standardization to obtain image quality levels at the lowest dose (Busch, 2004: 10).

Zhang *et al.* (2012: 236) explored the three-tier image quality concept with posterior anterior chest projections and noted that '*clinical-task-determined*' radiographic procedures offer radiation protection. Central to '*clinical-task-determined*' radiographic procedures are radiographers. Radiographers review the X-ray request form to assess if the patient's clinical history justifies the

X-ray request and to decide on the optimal imaging protocol. Therefore, paramount to exposure creep in digital X-ray imaging systems is the radiographer (Herrmann *et al.*, 2012: 1, 7; Moore *et al.*, 2012: 93).

CONCLUSION

This chapter has discussed conventional film screen radiography, CR, DR, EI, radiation protection, and exposure creep. The wide dynamic exposure latitude of digital X-ray imaging systems can disguise x-radiation overexposure; it is no longer a matter of a black image or a white image. X-radiation overexposure to patients may therefore go undetected. Given that the EI is valued as the only indicator of exposure technique in digital X-ray imaging systems, radiographers

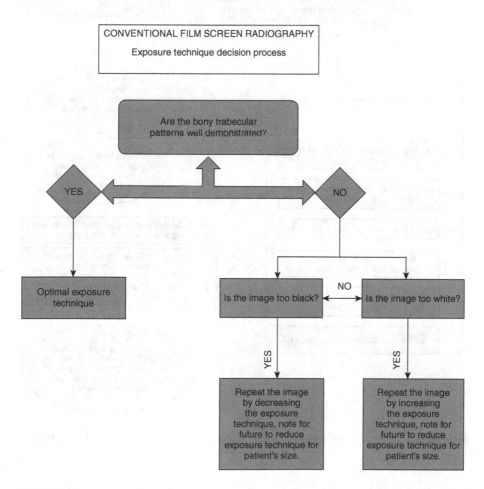

FIGURE 5.1 Conventional Film Screen Radiography Exposure Technique Decision Making Process

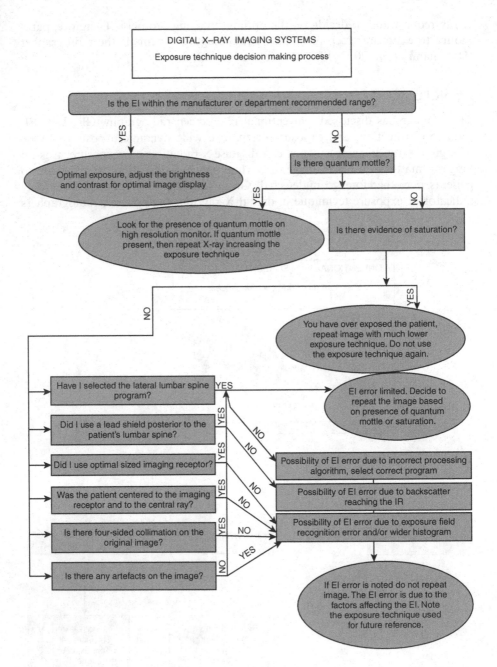

FIGURE 5.2   Digital X-ray Imaging System Exposure Technique Decision Making Process

need to use and understand EIs to halt this x-radiation overexposure. Yet, radiographers face challenges when using EI.

Varied manufacturer EI standards and names pose difficulties to radiographers working with equipment from different manufacturers. Having to learn the name of the EI and the optimal range for the equipment from different manufacturers may hinder EI compliance. There are efforts to standardize the EI, yet it remains unclear if it has been adopted worldwide. The factors affecting the EI need to be understood to decide correct exposure technique.

In digital X-ray imaging systems, exposure technique evaluation cannot begin with 'Does the image look okay? Is there optimal density (brightness) and contrast? Is the bony trabecular pattern and soft tissue optimally visualized?' It needs to start with: 'Is the EI within the recommended range?' This change in analyzing images needs to begin from the first year of radiography education. Radiography textbooks need to include the question: 'Is the EI within the recommended range?' as an evaluation criterion for exposure technique. Clinical practice will need to echo this change in order to shift the exposure evaluation criteria from the analog to digital era.

Comparing the exposure technique decision making process for analogue radiography (Figure 5.1) and digital X-ray imaging systems (Figure 5.2) for a lateral lumbar spine it becomes clear that EI is not black and white but has many grey areas to consider.

REFERENCES

Adam, A., Dixon, A.K., Gillard, J.H. and Schaefer-Prokop, C.M. (Editors). (2015). *Grainger & Allison's Diagnostic Radiology. A Textbook of Medical Imaging.* Edinburgh: Churchill Livingstone Elsevier.

Adambounou, K., Achy, O.B., Fiagan, Y.A., Adigo, A.M., Mondé, K., Gbande, P., Sonhaye, L., Tapsoba, T.L. and Adjenou, K.V. (2015). Knowledge and attitude of Togolese radiographers on medical irradiation of pregnant women. *Journal of Nuclear Medicine and Radiation Therapy,* 87(1): 1–5. doi:10.4172/2155-9619.1000S7-003.

Agfa-Gevaert. (1998). *ADC Compact Dose Monitoring Software User Manual.* Mortsel, Belgium: Agfa-Gevaert.

American Association of Physicists in Medicine (AAPM). (2009). *An Exposure Indicator for Digital Radiography.* Report of AAPM task group #116. Available from: www.aapm.org/pubs/reports/rpt_116.pdf

American College of Radiology (ACR), AAPM and Society for Imaging Informatics in Medicine (SIIM). (2014). ACR-AAPM-SIIM practice parameter for digital radiography. *Resolution 39 Amended 2014.* Available from: https://www.acr.org/-/media/ACR/Files/Practice-Parameters/Rad-Digital.pdf [Accessed 12 March 2020]

Artz, D.S. (1997). Computed radiography for the radiological technologist. *Seminars in Roentgenology,* 32(1): 12–24.

Bansal, G.J. (2006). Digital radiography. A comparison with modern conventional imaging. *Postgraduate Medical Journal,* 82(969): 425–428. doi:10.1136/pgmj.2005.038448.

Behling, R. (2016). *Modern Diagnostic X-Ray Sources. Technology, Manufacturing, Reliability.* Boca Raton, FL: CRC Press.

Brake, D.A. (2016). A standardized exposure index for digital radiography. *Radiologic Technology,* 87(5): 581–585.

Busch, H.P. (2004). *European Commission DIMOND III Image Quality and Dose Management for Digital Radiography Final Report*. Available from: www.sorf.fi/doc/diamond_III.pdf

Bushong, S.C. (2013). *Radiologic Science for Technologists Physics, Biology and Protection. 10th ed.* MO: Elsevier.

Campbell, S.S., Morton, D. and Grobler, A.D. (2019). Transitioning from analogue to digital imaging: Challenges of South African analogue-trained radiographers. *Radiography*, 25(2019): E39–e44. doi:10.1016/j.radi.2018.10.001.

Campeau, F.E. and Fleitz, J. (2017). *Limited Radiography. 4th ed.* Boston, MA: Cengage Learning.

Carlton, R.R. and Adler, A.M. (2013). *Radiographic Imaging Concepts and Principles. 5th ed.* New York: Delmar Cengage Learning.

Carrol, Q.B. (2014). *Radiography in the Digital Age. 2nd ed.* IL: Charles C. Thomas.

Carter, C. and Veale, B. (2014). *Digital Radiography and PACS. 2nd ed.* MO: Elsevier.

Cohen, M.D., Cooper, M.L., Piersall, K. and Apgar, B.K. (2011). Quality assurance: Using the exposure index and the deviation index to monitor radiation exposure for portable chest radiographs in neonates. *Pediatric Radiology*, 41: 592–601. doi:10.1007/s00247-010-1951-9.

Cohen, M.D., Markowitz, R., Hill, J., Huda, W., Babyn, P. and Apgar, B. (2012). Quality assurance: A comparison study of radiographic exposure for neonatal chest radiographs at 4 academic hospitals. *Pediatric Radiology*, 41: 592–601. doi:10.1007/s00247-010-1951-9.

Costa, A.M. and Pelegrino, M.S. (2014). Evaluation of entrance surface air kerma from exposure index in computed radiography. *Radiation Physics and Chemistry*, 104 (2014): 198–200. doi:10.1016/j.radphyschem.2014.05.005.

Curry, T.S., Dowdey, J.E. and Murry, R.C. (1990). *Christensen's Physics of Diagnostic Radiology. 4th ed.* Philadelphia, PA: Lea & Febiger.

Don, S., Whiting, B.R., Rutz, L.J. and Apgar, B.K. (2012). New exposure indicators for digital radiography simplified for radiologists and technologists. *American Journal of Radiology*, 199: 1337–1341. doi:10.2214/AJR.12.8678.

ECRI Institute. (2014). *Top 10 Health Technology Hazards for 2015*. Available from: www.ecri.org/Documents/White_papers/Top_10_2015.pdf

Fauber, T.L. (2017). *Radiographic Imaging and Exposure. 5th ed.* MO: Elsevier.

Gallet, J. (2010). *The Concept of Exposure Index for Carestream Directview Systems.* USA: Carestream Health.

Gibson, D.J. and Davidson, R.A. (2012). Exposure creep in computed radiography: A longitudinal study. *Academic Radiology*, 19(4): 458.

Guerra, A.D. (2004). *Ionising Radiation Detectors for Medical Imaging*. NJ: World Scientific.

Gupta, A.K., Chowdhury, V. and Khandelwal, N. (Editors). (2013). *Diagnostic Radiology Recent Advances and Applied Physics in Imaging. 2nd ed.* New Delhi: Jaypee Brothers Medical Publishers.

Haus, A.G. and Cullinan, J.E. (1989). Screen film processing systems for medical radiography: A historical review. *Radiographics*, 9(6): 1203–1224.

Hayre, C.M., Blackman, S., Eyden, A. and Carlton, K. (2019). The use of digital side markers (DSMs) and cropping in digital radiography. *Journal of Medical Imaging and Radiation Sciences*, 50(2019): 234–242. doi:10.1016/j.jmir.2018.11.001.

Herrmann, T.L., Fauber, T.L., Gill, J., Hoffman, C., Orth, D.K., Peterson, P.A., Prouty, R.R., Woodward, A.P. and Odle, T.G. (2012). *White Paper: Best Practices in Digital Radiography.* American Society of Radiologic Technologists. Available from: www.asrt.org/docs/default-source/whitepapers/asrt12_bstpracdigradwhp_final.pdf

Huda, W. and Abrahams, R.B. (2015). Radiographic techniques, contrast and noise in X-ray imaging. *American Journal of Roentgenology*, 204(2): 126–131. doi:10.2214/AJR.14.13116.

International Atomic Energy Agency (IAEA). (2011). *Avoidance of Unnecessary Dose to Patients While Transitioning from Analogue to Digital Radiology.* Vienna: IAEA.

International Atomic Energy Agency (IAEA). (2016). *Medical Exposure in Radiology: Optimization of Protection. Module VIII.3-Part 3: Operational Considerations.* Available from: http://slideplayer.com/slide/9317273/

International Electrotechnical Commission (IEC). (2008). *International Standard 62494-1: Medical Electrical Equipment- Exposure Index of Digital X-Ray Imaging Systems.* Geneva: IEC.

International Society of Radiographers and Radiological Technologists (ISRRT). (2018a). *ISRRT Response.* Available from: www.isrrt.org/isrrt-response

International Society of Radiographers and Radiological Technologists (ISRRT). (2018b). *Justification- Flowchart.* Available from: www.isrrt.org/justification-flowchart

Jeon, S. (2014). *Detector Exposure Indicator in GE X-Ray Systems.* Available from: www3.gehealthcare.ru/~/media/documents/russia/products/radiography/brochure/fixed/gehc_radiography-discovery-xr656_detector-exposure-indicator.pdf?Parent=%7BB5EB9550-6D39-4C3A-BB3B-C84C1F67A8B2%7D

Johnston, J.N. and Fauber, T.L. (2016). *Essentials of Radiographic Physics and Imaging. 2nd ed.* MO: Elsevier.

Jones, A.K., Heintz, P., Geiser, W., Goldman, L. Jerijan, K., Martin, M., Peck, D., Pfeiffer, D., Ranger, N. and Yorkston, J. (2015). Ongoing quality control in digital radiography: Report of AAPM Imaging Physics Committee Task Group 15. *Medical Physics*, 42(11): 6658–6670.

Khotle, T., de Vos, H., Herbst, C.P. and Rae, W.I.D. (2009). Optimization of exposure factors and image quality for computed radiography. In *World Congress on Medical Physics and Biomedical Engineering.* Edited by Dossel, O. and Schlegel. W.C. Germany: Springer.

Kim, D.S. and Kim, E. (2016). Noise power spectrum of the fixed pattern noise in digital radiography detectors. *Medical Physics*, 43(6): 2765–2775. doi:10.1118/1.4948691.

Koen, L., Herbst, C. and Rae, W. (2008). Computed radiography exposure indices in mammography. *SA Journal of Radiology*, 12(2): 28–30. doi:10.4102/sajr.v12i2.565.

Konica Minolta Medical and Graphic. (2010). *Direct Digitizer Regius Model 210 Technical Commentary/Image Adjustment Manual.* (Manual). Netherlands: Konica Minolta Medical & Graphic Imaging Europe B.V.

Körner, M., Weber, C.H., Wirth, S., Pfeifer, K., Reiser, M.F. and Treitl, M. (2007). Advances in digital radiography: Physical principles and system overview. *Radiographics*, 27: 675–686.

Kouamé, N., Ngoan-Domoua, A.M., Sétchéou, A., Nezou, B.J.P., Konan, K.D., N'Gbesso, R.D. and Kéita, A.K. (2012). Pregnancy and radiation risks in radiology: Users' knowledge at the University Hospital of Yopougon (Abidjan, Côte d'Ivoire). *Radioprotection*, 47(4): 553–560.

Kweon, D.C., Chung, W.K., Dong, K.R., Lee, J.W., Choi, J.W., Goo, E.H., Lee, J.S., Kim, S.G., Cho, J.H. and Chung, J.E. (2012). Evaluation of the radiation dose to a phantom for various X-ray exposure factors performed using the dose area product in digital radiography. *Radiation Effects & Defects in Solids*, 167(12): 954–970. doi:10.1080/10420150.2012.684060.

Lampignano, J.P. and Kendrick, L.E. (2018). *Bontrager's Textbook of Radiographic Positioning and Related Anatomy. 9th ed.* MO: Elsevier.

Lanca, L. and Silva, A. (2008). Evaluation of exposure index (lgM) in orthopaedic radiography. *Radiation Protection Dosimetry*, 2008(129): 112–118. doi:10.1093/rpd/ncn143.

Lanca, L. and Silva, A. (2013). *Digital Imaging Systems for Plain Radiography.* New York: Springer.

Lee, K.H., Kwon, J.W., Yoon, Y.C., Choi, S., Jung, J.Y., Kim, J.H. and Lee, S.J. (2009). Slot-scan digital radiography of the lower extremities: A comparison to computed radiography with respect to image quality and radiation dose. *Korean Journal of Radiology*, 10(1): 51–57.

Lewis, S., Pieterse, T. and Lawrence, H.A. (2019a). Evaluating the use of exposure indicators in digital x-ray imaging system: Gauteng South Africa. *Radiography*, In Press. doi:10.1016/j.radi.2019.01.003.

Lewis, S., Pieterse, T. and Lawrence, H.A. (2019b). Retrospective evaluation of exposure indicators: A pilot study of exposure technique in digital radiography. *Journal of Medical Radiation Sciences*, 66(2019): 38–43. doi:10.1002/jmrs.317.

Long, B.W., Frank, D.E. and Ehrlich, R.A. (2017). *Radiography Essentials for Limited Practice. 5th ed.* MO: Elsevier.

Luckey, G.W. (1975, January 7). *US Patent No. 3,859,527.* Washington, DC: U.S. Patent and Trademark office.

Martensen, K.M. (2015). *Radiographic Image Analysis. 4th ed.* MO: Elsevier Saunders.

Mc Fadden, S., Roding, T., de Vries, G., Benwell, M., Bijwaard, H. and Scheurleer, J. (2018). Digital imaging and radiographic practice in diagnostic radiography: An overview of current knowledge and practice in Europe. *Radiography*, 24(2018): 137–141. doi:10.1016/j.radi.2017.11.004.

McQuille-Martensen, K. (2014). *Radiographic Imaging Analysis. 2nd ed.* USA: Content Technologies.

Mojiri, M. and Moghimbeigi, A. (2011). Awareness and attitude of radiographers toward radiation protection. *Journal of Paramedical Sciences*, 2(4): 1–5.

Moore, Q.T., Don, S., Goske, M.J., Strauss, K.J., Cohen, M., Herrmann, T., MacDougall, R., Noble, L., Morrison, G., John, S.D. and Lehman, L. (2012). Image gently: Using exposure indicators to improve pediatric digital radiography. [Special report]. *Radiologic Technology*, 84(1): 93–99. Available from: www.pedrad.org/Portals/6/Procedures/RADT12_SeptOct_v84i1_Q.Moore.pdf

Moore, R. (1980). Computed radiography. *Med Electron*, 11(2): 78–79.

Mothiram, U., Brennan, P.C., Lewis, S.J., Moran, B. and Robinson, J. (2014). Radiography exposure indices: A review. *Journal of Medical Radiation Sciences*, 61: 112–118. doi:10.1002/jmrs.49.

Mothiram, U., Brennan, P.C., Robinson, J., Lewis, S.J. and Moran, B. (2013). Retrospective evaluation of exposure index (EI) values from plain radiographs reveal important considerations for quality improvement. *Journal of Medical Radiation Sciences*, 60(4): 115–122. doi:10.1002/jmrs.25.

Neitzel, U., Maack, I. and Günther-Kohfahl, S. (1994). Image quality of a digital chest radiography system based on a selenium detector. *Medical Physics*, 21(4): 509–516. doi:10.1118/1.597389.

Ng, C.K.C. and Sun, Z. (2010). Development of an online automatic computed radiography dose data mining program: A preliminary study. *Computer Methods and Programs in Biomedicine*, 97(2010): 48–52. doi:10.1016/j.cmpb.2009.07.001.

Paolicchi, F., Miniati, F., Bastiani, L., Faggioni, L., Ciaramella, A., Creonti, I., Sottocornola, C., Dionisi, C. and Caramella, D. (2016). Assessment of radiation protection awareness and knowledge about radiological examination doses among Italian radiographers. *Insights Imaging*, 7(2016): 233–242. doi:10.1007/s13244-015-0445-6.

Pesce, M. (2016, September 6). *Understanding Radiology Exposure Indicators.* [log post] Available from: www.carestream.com/blog/2016/09/06/understanding-radiology-exposure-indicators/

Peters, S.E. and Brennan, P.C. (2002). Digital radiography: Are the manufacturers' settings too high? Optimisation of the Kodak digital radiography system with aid of the computed radiography dose index. *European Journal of Radiology*, 12(2002): 2381–2387. doi:10.1007/s00330-001-1230-0.

Proctor, D.B. and Adams, A.P. (2014). *Kinn's the Medical Assistant: An Applied Learning Approach.* MO: Elsevier.

Ribeiro, L.P., Ribeiro, R.P.P., Almeida, S., Rodrigues, A.F.C.L., Abrantes, K.B., Azevedo, J.P. and Pinheiro, O. (2016). Exposure index in digital radiology. *Insights Imaging*, 7(Suppl): S162–S465. doi:10.1007/s13244-016-0475-8.

Rowlands, J.A. (2002). The physics of computed radiography. *Physics in Medicine and Biology*, 47(2002): 123–166.

Samei, E. (2003). Performance of digital radiographic detectors: Quantification and assessment methods. *RSNA Categorical Course in Diagnostic Radiology Physics*, 2003: 37–38.

Sandridge, T.M. (2017). Technical factors and exposure indicators. *Radiologic Technology*, 88(5): 572–573.

Sankaran, A. (1997). Computed Radiography (CR) using storage phosphor plate for a filmless radiology department – Recent trends (Update). In *Luminescence and Its Applications -97*. Edited by Bhushan, S. and Dewangan, P.K. New Delhi: Allied.

Schaetzing, R. (2003). *Advances in Digital Radiography: RSNA Categorical Course in Diagnostic Radiology Physics.* Available from: www.umich.edu/~ners580/nersbioe_481/lectures/pdfs/RSNA2003_CR_Schaetzing.pdf

Scott, A.W., Zhou, Y., Allahverdian, J., Nute, J.L. and Lee, C. (2016). Evaluation of digital radiography practice using exposure index tracking. *Journal of Applied Clinical Medical Physics*, 17(6): 343–355.

Seeram, E. (2019). *Digital Radiography Physical Principles and Quality Control. 2nd ed.* Singapore: Springer.

Seeram, E. and Brennan, P.C. (2017). *Radiation Protection in Diagnostic X-Ray Imaging.* USA: Jones & Bartlett Learning.

Seeram, E., Davidson, R., Bushong, S. and Swan, H. (2013). Radiation dose optimization research: Exposure technique approaches in CR imaging – A literature review. *Radiography*, 19(2013): 331–338.

Seeram, E., Davidson, R., Bushong, S. and Swan, H. (2016). Optimizing the exposure indicator as a dose management strategy in computed radiography. *Radiologic Technology*, 87(4): 380–391.

Seibert, J.A. and Morin, R.L. (2011). The standardized exposure index for digital radiography: An opportunity for optimization of radiation dose to the pediatric population. *Pediatric Radiology*, 41(5): 573–581. doi:10.1007/s00247-010-1954-6.

Shah, A.S., Begum, N., Nasreen, S. and Khan, A. (2007). Assessment of radiation protection awareness levels in medical radiation science technologists – A pilot survey. *Journal of Postgraduate Medical Institute*, 21(03): 169–172.

Shepard, J.S., Flynn, M., Gingold, E., Goldman, L., Krugh, K., Leong, D.L., Mah, E., Ogden, K., Peck, D., Samei, E., Wang, J. and Willis, C.E. (2009). An exposure indicator for digital radiography: AAPM Task Group 116 (executive summary). *Medical Physics*, 36(7): 2898–2914. doi:10.1118/1.3121505.

Sherer, M.A.S., Visconti, P.J., Ritenour, E.R. and Haynes, K.W. (2018). *Radiation Protection in Medical Radiography. 8th ed.* St Louis, MO: Elsevier.

Siegel, E.L. and Kolodner, R.M. (Editors). (2006). *Filmless Radiology.* USA: Springer.

Singh, H., Sasane, A. and Lodha, R. (2016). *Textbook of Radiology Physics.* New Delhi: Jaypee Brothers Medical.

Suetens, P. (2002). *Fundamentals of Medical Imaging.* United Kingdom: Cambridge University Press.

Szarmach, A., Piskunowicz, M., Swieton, D., Muc, A., Mockatto, G., Dzierzanowski, J. and Szurowska, E. (2015). Radiation safety awareness among medical staff. *Polish Journal of Radiology*, 2015(80): 57–61. doi:10.12659/PJR.892758.

Takaki, T., Takeda, K., Murakami, S., Ogawa, H., Ogawa, M. and Sakamoto, M. (2016). Evaluation of the effects of subject thickness on the exposure index in digital radiography. *Radiology Physics Technology*, 9(2016): 116–120. doi:10.1007/s12194-015-0341-2.

Takakia, T., Fujibuchic, T., Murakamib, S., Aokid, T. and Ohkic, M. (2019). The clinical significance of modifying X-ray tube current-time product based on prior image deviation index for digital radiography. *Physica Medica*, 63(2019): 35–40. doi:10.1016/j.ejmp.2019.05.011.

Talab, A.H.D., Mahmodi, F., Aghaei, H., Jodaki, L. and Ganji, D. (2016). Evaluation the effect of individual and demographic factors on awareness, attitude and performance of radiographers regarding principles of radiation protection. *Al Ameen Journal of Medical Science*, 9(2): 90–95.

Thomas, A.M.K. and Banerjee, A.K. (2013). *The History of Radiology*. Oxford: Oxford University Press.

Tighe, M.M. and Brown, M. (2015). *Mosby's Comprehensive Review for Veterinary Technicians. 4th ed.* MO: Elsevier.

Torigian, D.A. and Ramchandani, P. (2017). *Radiology Secrets Plus. 4th ed.* Philadelphia, PA: Elsevier.

Trapnell, D.H. (1967). *Principles of X-Ray Diagnosis*. London: Butterworths.

Uffmann, M. and Schaefer-Prokop, C. (2009). Digital radiography: The balance between image quality and required radiation dose. *European Journal of Radiology*, 72 (2009): 202–208.

Vastagh, S. (2011). Statement by MITA on behalf of the MITA CR-DR group of the X-ray section. *Pediatric Radiology*, 41(5): 566–567. doi:10.1007/s00247-010-1961-7.

Verma, B.S. and Indrajit, I.K. (2008). Impact of computers in radiography: The advent of digital radiography, Part-2. *Indian Journal of Radiology and Imaging*, 18(3): 204–209. doi:10.4103/0971-3026.41828.

Vyborny, C.J. (1997). Image quality and the clinical radiographic examination. *Radiographics*, 17(2): 479–498. doi:10.1148/radiographics.17.2.9084085.

Warren-Forward, H, Arthur, L., Hobson, L., Skinner, R., Watts, A., Clapham, K., Lou, D. and Cook, A. (2007). An assessment of exposure indices in computed radiography for the posterior-anterior chest and the lateral lumbar spine. *The British Journal of Radiology*, 80(949): 26–31.

Washington, C.M. and Leaver, D. (2016). *Principles and Practice of Radiation Therapy. 4th ed.* MO: Elsevier Mosby.

Whitley, S., Jefferson, G., Holmes, K., Sloane, C., Anderson, C. and Hoadley, G. (2016). *Clark's Positioning in Radiography. 13th ed.* London: CRC Press.

Williams, M.B., Krupinski, E.A., Strauss, K.J., Breeden, W.K., Rzeszotarski, M.S., Applegate, K., Wyatt, M., Bjork, S. and Seibert, J.A. (2007). Digital radiography image quality: Image acquisition. *Journal of the American College of Radiology*, 4(6): 371–388.

World Health Organisation (WHO). (2012a). *Bonn Call – For-Action Joint Position Statement by the IAEA and WHO*. Available from: www.who.int/ionising_radiation/medical_exposure/Bonn_call_action.pdf

Zhang, M., Zhoa, B., Wang, Y., Chen, W. and Hou, L. (2012). Dose optimization for different medical imaging tasks from exposure index, exposure control factor, and MAS in digital radiography. *Health Physics*, 103(3): 235–240. doi:10.1097/HP.0b013e31824e71b6.

Section 3

Patient Care and Considerations

6 Patient-Centered Care and Considerations

Emma Hyde and Maryann Hardy

INTRODUCTION

In this chapter, we will look at the origin and development of patient-centered care. We will consider the nature of the profession of diagnostic radiography and the ways that diagnostic radiographers can deliver patient-centered care. The purpose of the patient's attendance, which in most cases will be the production of diagnostic images to aid diagnosis, and the use of radiation means that this can be challenging, particularly as technical competence is essential for assuring patient safety and satisfactory examination outcomes. Alongside this, the pressure to work efficiently and maximize the use of imaging equipment means that radiographers are expected to image many patients during the course of a single shift. Consequently, it is crucial that radiographers are able to balance safety and efficiency with the needs of patients.

The learning outcomes for this chapter are as follows:

- to understand the development of patient-centered care in the UK and what good patient-centered care looks like;
- to think about how healthcare is changing, how radiographic practice is changing, and the impact this may have on patient-centered care;
- to consider the role of the diagnostic radiographer in imaging examinations and procedures and how service efficiency might be maintained while ensuring patient-centered care.

EVOLUTION OF PATIENT-CENTERED CARE

Historically, patient care in the UK has been paternalistic, adopting a medical model approach. The medical model of care is often referred to as a reductionist 'diagnose-treat-discharge' process with care focused on the anatomy or body system of the presenting complaint with limited regard for the whole person and the wider health and social care needs they may have. This approach has evolved predominantly from the way healthcare services have been organized, funded, and subsequently delivered by the UK National Health Service (NHS). The organizational structure of the

NHS revolves around three identifiable tiers: primary care in the community setting; secondary care for services within the hospital setting; and tertiary care for highly specialized hospital care for more serious or complex conditions. While it is expected that patients self-refer themselves to their general practitioner or family doctor initially, further diagnostic tests and/or treatment require the medical professional to refer the patient into the appropriate services, effectively acting as the gatekeeper to care, fostering a culture of 'doctor knows best'. The doctor–patient relationship has been defined by White (2006) as an:

> ... interaction that changes over the course of [an] illness and depending on the symptoms. Thus, it may take the form of an activity (doctor) passivity (patient) in the context of intensive care; guidance-cooperation, where the treatment is of an acute illness; and mutual participation in terms of managing on-going illness.

This definition highlights the traditional dominance of the medical doctor within the care relationship, particularly where illnesses are acute or short-lived, with the patient acting as a passive recipient of care rather than equal partner.

This paternalistic approach to patient care was exacerbated further with the widespread adoption of evidence-based medicine in the 1980s, which advocated that medical decision-making should be based upon research evidence which readily underpinned the growing library of treatment guidelines. While Sackett et al. (1996) argued that evidence-based medicine was not a 'one size fits all' cookbook approach to medicine, but instead combines clinical expertise with patient values and expectations as well as research evidence to inform clinical decision-making, these components are not equal and difficulties still exist in reconciling evidence-based medicine with the more humanistic patient-centered care model (Weaver, 2015).

At a similar time to the emergence of evidence-based medicine, the Picker Institute (a not-for-profit organization) was established to develop and promote patient-centered care, with the vision of the highest quality person centred care for all, always. The Picker Institute remains dedicated to researching how healthcare organizations can improve patient experience, while ensuring patient's clinical needs are met (Picker Institute Europe, 2019). They have developed eight principles of good person-centered care, which are the following:

* fast access to reliable healthcare advice;
* effective treatment delivered by trusted professionals;
* continuity of care and smooth transitions;
* involvement of, and support for, family and carers;
* clear information, communication, and support for self-care;
* involvement in decisions and respect for patient's preferences;
* emotional support, empathy, and respect;
* attention to physical and environmental needs.

Importantly, over the period from organizational instigation to publication of their eight principles, the language used by the Picker Institute changed from 'patient' to 'person' signifying a change in the underpinning ethos of health-care delivery. Williamson et al. (2009) and others have argued that the use of the term 'patient' is disempowering, fostering dependence and subservience to the medical profession and advocate that terms such as client or service user are more empowering. The medical profession itself is also starting to question the nature of their role, considering how they may need to adapt to meet the healthcare needs of a population with a longer life expectancy and subsequently, an increase in long-term health conditions and comorbidities (Iacobuci, 2018).

The patient-centered care movement is not limited to the UK but instead is a global phenomenon repositioning the patient to the center of all healthcare decisions and leading to a change in the delivery of services. In the UK, this changing perception has resulted in a radical rethink of the NHS ethos. Publications such as the recent 'NHS 10 Year Plan' (2019) and the Nuffield Trust's 'Shifting the balance of care' (2017) clearly argue for more holistic models of healthcare that consider the whole person, not just the presenting symptom.

While the original principles of person-centered care developed by the Picker Institute provide insight into the expectations of person-centered care from the perspective of care users or patients, one of the UK's leading health think tanks, The Health Foundation (2014), defines the organizational delivery of patient-centered care as follows:

- affording people dignity, compassion, and respect;
- offering coordinated care, support, or treatment;
- offering personalized care, support, or treatment;
- supporting people to recognize and develop their own strengths and abilities to enable them to live an independent and fulfilling life.

These four key areas are also embedded within the person-centered approaches framework set out by Health Education England (2016) and inform NHS 'Sustainability & Transformational Plans' in England (2017) emphasizing the centrality of patient-centered care within health policy and care delivery in the UK today.

However, a change in human perspective is not the only revolution occurring in healthcare. With increasing computer power and greater technological automation, healthcare is simultaneously experiencing a technological revolution. Nowhere is this more apparent than within radiology where imaging technologies are unrecognizable from those in common use in the 1980s with the introduction of digital radiography (DR), hybrid imaging technologies, advancements in cross-sectional imaging technologies, and the rise of artificial intelligence providing opportunities to incorporate genomics, radiomics, and other 'big data' analytics into diagnostic decision-making and clinical interpretation. Without doubt, these technological developments are fundamentally changing healthcare delivery and will continue to do so into the future as we adapt our care approaches to meet the needs of an increasingly technology-enabled society. The growing importance of technology in healthcare was highlighted in the Topol report (2019), which set out a vision for how technology will change healthcare between now and 2040, with a push toward more primary care based services appearing inevitable. In addition, advancing technology will lead to changing roles and responsibilities with an increase in healthcare professionals such as nurses, radiographers, and occupational therapists moving into advanced practice roles and taking on tasks previously within the domain of medical practitioners. Topol (2019) suggests that the increased capacity provided by advanced technological solutions should facilitate the creation of more joined-up care pathways with practitioners able to focus more on delivering care than on the processes of care themselves. Topol also outlines how technology adoption could improve communication, provide greater access to relevant information, and fundamentally, more choice in care delivery, all of which are core components of patient-centered care (Health Foundation, 2014).

Despite the future vision in the Topol report outlining the enabling possibilities of technology, within diagnostic imaging, digital technologies have been embedded without true consideration of their impact on the working practices of staff or the patient experience. It is evident to those who have observed the transition from film-screen to digital imaging that the introduction of DR in imaging departments has changed the both the opportunity for, and content of, interactions between diagnostic radiographers and patients. DR has made the acquisition and processing of imaging examinations quicker. However, this increase in speed brings with it challenges. The shorter acquisition and image processing time has led to an expectation of shorter examination times and a push to maximize efficiency and increase patient throughput volumes. Efficiency savings have also been encouraged through the employment of healthcare assistants to undertake many patient-facing roles. While the increase in patient throughput can be seen as beneficial in terms of reduced waiting times, the delegation of patient-facing roles to assistants may also impact on the radiographers' ability to provide patient-centered care. For example, previously the radiographer may have been involved in getting patients changed for their examination whereas today this is often delegated to an imaging assistant. While this appears to be a simple task for delegation, the opportunity for the radiographer to meet the patient in the waiting room and observe and communicate with them on the journey to the changing area permitted the radiographer to undertake an initial assessment of the patient's physical and mental capabilities prior to the examination. It also provided an opportunity to prepare the environment, ensure appropriate manual handling aids and additional imaging aids were available, and plan for any necessary technique modification. Now, this assessment often takes place with the patient already in the imaging room and the patient may already be on the examination table before it is realized that additional sponge pads, sandbags, etc. are required. This can lead to radiographers appearing unprepared, searching around for items and create doubt in the patients' mind of the radiographer's technical competence – not the ideal scenario for patient-centered care.

In addition to the limited radiographer preparation time, lone working and almost instantaneous within-room image processing that modern DR systems are able to undertake means that the opportunity to seek opinions, advice, and support from colleagues has been reduced. The act of taking imaging cassettes to a processor, waiting for films to process, and viewing images within a collaborative central space provided radiographers with time to discuss techniques and/or approaches to supporting patients during their imaging examination. Modern departments rarely have this central 'gathering' area in their design and this means to ask advice, radiographers would need to leave the room, and the patient, to seek professional assistance or second opinion. Clearly this is not ideal, especially if the patient is particularly vulnerable or unwell, and so opportunities to ask for advice, to develop new technical skills, and benefit from peer learning may be missed. Consequently, within radiography, there is little evidence to support the theory that the move to DR has resulted in a better experience for patients or patients. Radiographers must consider how their practices, and changing technologies and departmental operation, impact on patient experience and balance the demands of technical efficiency with patient-centered care.

Radiography is a unique profession, which has short, focused interactions with patients. This brings challenges in terms of balancing efficiency and patient-centered care. In the next section, we will consider this in more detail and propose methods to support radiographers to deliver patient-centered care.

UNIQUENESS OF RADIOLOGY

The role of a diagnostic radiographer is unique within healthcare practice. Patients attend for short appointments (often single attendances) where the purpose is to acquire diagnostic images to enable appropriate care decisions to be made. This can result in radiographers focusing primarily on the technical aspect of the role, that is, image acquisition. This tendency is more common among student or newly qualified radiographers, who are still learning the technical skills necessary to be a competent diagnostic radiographer. This emphasis on technical success can result in a reduced focus on the person being imaged, and their needs, and consequently can be perceived as a lack of patient-centered care.

In order to redress this, we have outlined a number of patient-centered approaches below that maybe adopted and which are proposed based on research conducted by the authors in 2018–2019. While these approaches focus on adult patients, they are transferrable to children and young people.

COMMUNICATION

Effective communication with patients is the fundamental building block of delivering patient-centered care. Radiographers should be communicating clearly with patients' from the moment the patient arrives in the imaging department, starting with a statement of welcome and introduction. The 'Hello my name

is ...' campaign, launched by Dr Kate Granger, MBE, and her husband Chris
Pointer in 2013, aimed to embed a culture of compassion into healthcare, initiated by Kate's experiences after a cancer diagnosis. Although Kate passed
away in 2016, her legacy lives on and the campaign is now used by healthcare
professionals at all levels and recognized across the world as the first step in
improved patient care. In the UK, many NHS Trusts have purchased 'Hello
my name is ...' name badges for their staff, to help embed the practice within
the organization, and have actively encouraged all staff to use this simple
phrase of introduction (Hello my name is, 2020).

> 'Hello my name is ... [insert name] and I am the radiographer who will be under-
> taking your [insert imaging examination]' is a simple but effective way to start any
> imaging examination or procedure. It allows the radiographer to introduce them-
> selves, and their role, and confirm the imaging examination, before the examination
> begins. It allows the patient to get to know the radiographer, confirm their attend-
> ance purpose and build a degree of trust and confidence in their care by the radiog-
> rapher, before the examination starts. This phrase takes just a few seconds, but is
> incredibly effective in starting to build a relationship, however short lived, with
> a patient. It builds trust and confidence, and helps to reduce the vulnerability that
> patients may be feeling.

Importantly, communication with patients should be ongoing throughout the
imaging examination or procedure, offering support and encouragement to the
patient at all times. After all, while seeing a patient laying on a hard table in
a hospital gown with arms and legs in specified positions may be routine for
the radiographer, it may be the first experience of imaging for the patient and
leave them with uncertainties and doubts. As a result, the radiographer should
ensure that the language used is appropriate to the understanding of the
person being cared for, and medical terminology should generally be avoided.
Clarification of the length of time an examination or procedure might take
can be reassuring, particularly in case of lengthier examinations or where the
patient is in discomfort, as this may ease anxieties such as 'they've been at it
a long time, they must have seen something' and support examination or positioning compliance. Care should be taken to be accurate about examination
timings, as patients may become frustrated or upset if they feel that their
imaging examination is taking longer than expected. This can also lead to
concern over the radiographer's technical or procedural competence. Ongoing
communication also provides opportunities for patients to ask questions that
are important to them, and these should be answered honestly, directing the
patient to where the answer might be found if unable to answer questions directly. Specifically, patients often have questions about where and when they
might receive results and any necessary information at the end of the examination or procedure. Clear instructions should be given, in simple language.
Where there are any doubts about understanding, the radiographer should
confirm the verbally communicated information with the patient or provide
the information in a written format for the patient to take home.

It is important to acknowledge the impact of nonverbal communication as well verbal communication on the patient–radiographer relationship. The impact of a smile, the use of appropriate touch (for example a gentle hand on the shoulder), good eye contact, and adopting an open posture should not be underestimated. These little gestures can make a huge difference to an anxious patient, and do not take any extra time. Lack of eye contact and a closed posture (such as folded arms or working with your back to the patient) can quickly make a patient feel uncomfortable, and that they are an inconvenience, particularly if they have physical needs that restrict ease of movement. Radiographers should reflect on and consider their nonverbal communication at all times, to ensure they are being perceived as open, friendly and reassuring to patients.

In summary, good communication is key in achieving a successful, patient-centered imaging examination or procedure. It is essential to ensure communication is clear, using simple language, and accompanied by consistent, supportive nonverbal communication. Poor communication may lead to a distressing or confusing experience for patients, particularly where the examination procedure and their role within it are unclear. A clear explanation of the procedure and ensuring patient understanding are paramount to good quality, patient-centered care. While explanation maybe considered a facet of communication, its importance is such that it has been considered separately in the next section.

EXPLANATION

A thorough explanation of the imaging examination or procedure, with pauses to allow patients to ask questions, is essential. It can be easy to fall into a routine of using the same 'script', which may not be suitable for all patients, depending on their previous knowledge and experience of imaging, and may not answer their questions and concerns. Of course, it is important to recognize that patients who have had a particular imaging examination or procedure, or regularly experience imaging to monitor treatment or follow-up progression of their condition, often will not need as much explanation or information as someone who is having the imaging examination or procedure for the first time. However, it is likely that most patients will have questions about what is going to happen to them and how long it will take, and where and when results will be available. It is also worth remembering that diagnostic imaging environment and the equipment can be daunting for patients, and with the requirement to discuss and ensure patient's understanding of the risks and benefits of a medical exposure to ionizing radiation prior to examination (under the IR(ME) regulations 2017), patients may have further anxieties over potential harm. Spending a few minutes explaining the examination or procedure is time well spent, and will often reduce the likelihood of patients becoming distressed or unable to tolerate the procedure.

Many imaging examinations also require patient safety checks prior to commencement, such as an MRI safety or a contrast media questionnaire, to ensure it is safe for the patient to undergo a particular test. These safety checks should be carried out before patients are asked to change their clothing, as if any safety issues arise, the imaging may not be able to proceed, so the patient will have undressed and changed unnecessarily. Importantly, radiographers should be available to answer any questions while completing safety questionnaires, as the reasons for, and purpose of, the questions may not be clear and may result in inaccurate responses, compromising patient safety during the examination. Examples of this may be not disclosing allergies, not being accurate with the date of last menstrual period, or failing to declare metal artifacts or devices, etc. Clear explanations of why the safety questions are being asked help to reduce this.

It is also important to clearly explain what the changing requirements are for the imaging examination. Being clear about which items of clothing to take off and why, and what clothing may be kept on, is key to avoiding confusion and potential patient embarrassment. It is also essential that clear instructions are provided on what items of clothing to put on for the test (such as a hospital or X-ray gown), and how to do this (which way any ties or slits might go) should also be clear. Patients can mistakenly put X-ray gowns on back to front, leading to privacy and dignity being compromised and potential embarrassment for patients and any observers. X-ray gowns with ties can be particularly difficult to fasten at the back as intended for patients with reduced mobility and are often left open or worn back to front with the ties fastened at the front, which might impact on both patient dignity and confidence. As a result of this, more departments are moving toward the use of theater-style trousers and tops (known as scrubs) rather than gowns, particularly for longer examinations such as MRI. In addition to explaining changing requirements, radiographers should consider whether patients need support to change, from either their carer or a member of the imaging department team (as appropriate).

Wherever possible, patients should be asked to change in an area which maintains privacy and dignity. Cubicles which are constructed from solid materials on all sides are preferable. Cubicles utilizing a curtain can cause anxiety and distress, particularly where they face into the waiting room or open corridor, due to the possibility of curtain movement. Ideally, changing cubicles should lead straight into the room where the patient will have their imaging examination, so that there is no need to return to sit in the waiting area once changed. Patients can find it embarrassing and undignified to sit in a public waiting area in a hospital gown, particularly if it is ill-fitting, and this should be avoided wherever possible. Further, observing someone in a hospital gown, particularly one that is ill-fitting or incorrectly fastened, can cause embarrassment and discomfort for other persons seated in the waiting area, and may impact on their perception of patient-centered care.

One patient anxiety that can be easily overlooked is that of security of personal belongings. It is important to clearly communicate where patients should put their belongings once they are changed. If available, secure lockers are the best option for keeping items safe. Bags or baskets in which patients can carry their belongings with them are a common option but these are not always suitable for patients with mobility problems or large enough to carry the many layers of clothes worn in the winter. As a result, radiographers should consider carefully the best arrangements for the security and transfer of personal belongings when assessing the patient on introduction prior to patient changing.

As well as explaining the arrangements for storage and transfer of personal belongings, radiographers should clearly explain whether or not personal belongings can be taken into the examination room. In some areas, such as MRI, this is not possible due to the risk of items becoming magnetized by the scanner and potentially becoming projectiles. In this environment, patient lockers should be available rather than personal belongings being left with the radiographer in the control room. In other parts of the department, patients may be permitted to take their belongings into the examination room, but a defined storage area should be identified for the patient to place their belongings to prevent them becoming a trip hazard if placed in random locations on the floor. Such forward thinking and careful environmental design are essential to prevent confusion, maintain safety, and ensure patients have confidence that their belongings are secure during their imaging examination.

Explaining clearly the procedure for patients to receive their results of their imaging examination or procedure is essential, and the radiographer should not assume that the patient understands who their referrer is or who will receive their results to make a decision about further treatment. There are several different ways that patients may receive their results. For patients attending as outpatients, the results will usually be sent back to the healthcare professional that referred them for their imaging (GP, specialist consultant, advanced practitioner, or other healthcare practitioner), and the radiographer should ensure the patient understands where to go and the time frame for the results being ready. For inpatients, the length of time between imaging being undertaken and the results being available will depend on the clinical urgency, but patients should be encouraged to speak to medical and nursing staff on their ward to expedite their care. Lack of clarity about the process of obtaining results can cause confusion and anxiety.

Alongside an explanation of where and when the results of the examination will be available, there is a need to provide a clear and transparent explanation of why the radiographer cannot provide the examination results. Some patients may be under the impression that the person undertaking their imaging can also interpret the images and give them the results. While this may be true where an immediate radiographer reporting service is in operation, this is not common practice across imaging modalities. Being evasive can raise patient's anxiety levels, and there can be a tendency to assume that something is very wrong. Providing a straightforward response to the patient explaining how the reporting process works and who is responsible for reporting the images will hopefully reassure them and reduce any anxiety.

Patients whose first language is not English should be offered interpreter services whenever possible. Ideally, space should be provided on the referral form to indicate that this is required. The use of family members as interpreters is not appropriate, as the patient may not wish for their family member to answer questions on their behalf, or indeed be fully informed about their condition. There is also uncertainty over the accuracy of the interpretation and response. This can be a particular issue if the family member is the child (minor) of the patient. The use of the professional interpreter services ensures dignity and respect for patients. If interpreter services are not available, for example, due to the patient having an urgent hospital admission, then online interpreter services such as Google Translate can be a useful alternative option.

A clear explanation of imaging examinations or procedures is essential to successful diagnostic imaging. Imaging departments can be a daunting place for patients, due to the unfamiliar equipment and potential risks involved. Spending time explaining the imaging examination or procedure often prevents issues with patients becoming distressed or unable to tolerate the imaging test, and increases the chances of a successful examination.

TAILORED CARE

Direct patient care within imaging departments commences from the time the patient arrives for their imaging examination. Consideration should be given to the design of, and space surrounding, the reception desk. While arrival in department maybe logged by electronic registration devices, many departments still use person-manned reception desks. Where this is the case, thought should be given to the information asked of the patient and the confidentiality of the surroundings. Not everyone will be comfortable divulging their name, address, telephone number, or examination details within earshot of other people.

Patients will often be directed to a waiting area prior to their examination, and this should offer a choice of seating to accommodate individual needs. For example, patients with a history of low back pain will find low, armless, soft-seating difficult to stand up from, and will usually prefer higher, slightly harder seating. Similarly, parents with young children may prefer low seats close to play equipment or books to allow ease of access to their child. As well as a choice of seat heights, a choice of seat widths is helpful for patients. Population obesity across the globe continues to be a health problem with many associated comorbidities, and bariatric patients may have some concerns over the comfort and stability of some chair designs. Wider widths will be more comfortable and reduce any anxiety or fear over getting stuck. While these points may seem simple, they are often overlooked and can make a real difference to a patient's experiences of imaging.

Patients with autism, dementia, or mental health conditions can find imaging department waiting areas distressing, particularly when they are busy or noisy. Consideration should be given to patient's needs, where these are known, at the time of booking the appointment to reduce stress and identify a quieter time slot where available. The use of subwaiting areas, or quieter areas of the department, might also be helpful to provide a calmer environment for such patients to wait in. This is particularly important if there is going to be a wait or delay before the patient's imaging is completed. Many imaging departments already have separate waiting areas for inpatients, out-patients, and emergency department patients. This separation is important to maintain dignity for very poorly or severely injured patients. This is also a sensible approach when imaging children. Providing a space specifically designed for children and their families, including thought for maneuvering or storing equipment, such as prams and pushchairs, is helpful. The use of calmer areas for patients with conditions such as dementia is less common, but one that is worth considering if there is an opportunity, and space, to create such an area.

Alongside the considerations of waiting environment and equipment, there is also the issue of clinical need. When presented with a very poorly or severely injured patient, a confused patient, or a parent with two or more children, radiographers should consider the clinical need of the patients currently waiting for their imaging examination or procedure and the staffing available in the department to assist with imaging. It may be appropriate to prioritize patients with additional care needs to ensure their medical care is expedited, or personal stress and anxiety levels are minimized. It is also appropriate to consider staff availability and prioritize patients who may need additional imaging staff support, particularly if their appointment is immediately before lunch breaks or toward the end of a working shift. It is not acceptable to merely image patients in order of arrival for while this may appear equitable, it does not take into consideration clinical need. This assessment of patient need and clinical judgement of prioritization needs to be undertaken on a case-by-case basis, and may require advice and support from senior colleagues in the team.

Maintaining patient's dignity during their imaging examination or procedure is a crucial part of patient-centered care. Offering a choice of radiolucent clothing to wear is a key way to maintain patient's dignity. A choice of gown size or style should be available to address the diversity of human sizes and shapes. Offering more substantive clothing such as theaters scrubs or tracksuits is useful for examinations with long examination times, where patients are at risk of becoming cold. The availability of a range of clothing options also helps to support patients with cultural requirements.

Linked to clothing is warmth. Imaging rooms are usually well air conditioned to ensure the optimal working temperature for the equipment. However, this often results in patients feeling cold. This may result from feeling unwell or being frail and sensitive to the cold, and specific care of patient warmth should be made during longer examinations such as angiography or MRI scanning. The provision of dressing gowns, blankets, duvets, or other items is essential in these situations to ensure patients feel warm and comfortable during their imaging. There should be a ready, clean supply of these items within the imaging department.

Offering a choice of lighting and music (if available) can be a useful way to tailor a patients experience and deliver patient-centered care. Patients experiencing visual disturbances may prefer lower lighting levels, and so room lights should be dimmed accordingly. Some patients may prefer brighter lighting so that they can see their surroundings more clearly, and again the room lights should be adjusted accordingly. The use of electronic 'window' panels on the ceiling for patients undergoing cross-sectional imaging or eye masks, where these are available, may also be helpful, particularly in MRI. The use of music can be reassuring and soothing and can be a helpful distraction that enables patients to tolerate any discomfort experienced during the imaging examination more easily. However, it is important to check the patient's musical taste and adapt the radio station, choice of CD, or online music streaming service accordingly.

Maintaining the position required for many imaging examinations or procedures can be hard, and patients, particularly those who are unwell, frail, or have a disability, can become fatigued with the effort or experience pain. The use of sponge pads, sandbags, or pillows is a good way to tailor care and support the patient to find a position that is comfortable for them, while ensuring correct positioning for the imaging acquisition. It is a crucial part of the explanation of the examination to ensure the patient understands the positioning requirements, and to ask them what will help them maintain the position. There are likely to be differences depending on age, mobility, and pain being experienced at the time of imaging, so radiographers should not make assumptions but instead involve the patient in the process. Alternative or modified techniques and patient positions or rest breaks should be considered wherever possible, particularly for longer examinations. This can make the difference between a patient being able to tolerate their imaging examination or procedure, or not.

There are many ways that radiographers can tailor imaging examinations or procedures to patients' needs and deliver patient-centered care. Some of these depend on the resources available in the department, such as gown sizes, differing height and width chairs, etc. However, with prior thought, there are still lots of ways to deliver tailored, patient-centered care while simultaneously ensuring clinical needs are met.

THE ROLE OF CARER

Some patients attend the imaging department accompanied by a carer. This may be a formal carer (e.g., nursing home staff) but will more often than not be a family member or friend. The role of the carer, unless they are a parent/guardian of a young child, is often overlooked by radiographers, although they can often be very helpful during the imaging examination or procedure. Of course, there are safety considerations, which need to be taken into account when carers enter the imaging room with the patient, but if these are addressed, the carer usually helps ensure a successful examination. A carer will usually know the patient's likes, dislikes, and how best to assist them.

It is important to discuss with both the patient and their carer what role the carer will take during the imaging procedure, so that there is a shared understanding of the carer's role. For example, the carer maybe a neighbor or a young relative of the patient, and therefore it may not be appropriate to ask the carer to help the patient to change, as this may lead to embarrassment for both patient and carer. However, if the carer is the partner or close family member of the patient, the carer may be used to providing support for the patient to change or mobilize and may wish to provide this support on the imaging department too. Each situation will be different, and assumptions should not be made as to the role each carer will undertake.

When including carer's in the imaging acquisition process, it must always be remembered that it is the patient that should be communicated with about what is going to happen. Just because there is a carer in attendance does not mean that the patient cannot comprehend instructions or communicate their wishes or respond to questions. This is also important if the patient is unconscious. There is a wealth of evidence that unconscious patients are aware of their surroundings and able to hear and remember being spoken too (Elliott & Wright, 1999). As such, communication should always be maintained with the patient, and this reinforced by use of the patient's name. If the carer needs to respond on the patient's behalf, then, of course, they can do this, but only if this does not contravene the rights and wishes of the patient. Remember, the responses from the carer maybe colored by their previous experience of the patient and their condition, and may not necessarily represent the perspective of the patient themselves.

Without doubt, the inclusion of the carer in the imaging examination can make the experience more patient centered and aid the radiographer, but careful consideration of the role of the carer is crucial to ensure unnecessary embarrassment is avoided.

TIME

Patients are often aware of the time pressures that radiographers are working under. Media stories frequently report hospital pressures and staff shortages, and as a result patients may observe a busy waiting room, and feel obliged to take up as little time as possible during their imaging examination or procedure. This may result in the examination commencing and patients realizing they do not fully understand what is going to happen, making the experience very distressing. Alternatively, patients may leave the Imaging Department before realizing that they are unclear about what happens next and how to get their results. It is therefore the role of the radiographer to ensure that patients do not feel rushed or under pressure during their imaging, even when the department is busy.

This does not mean that the radiographer should not work efficiently, and efficient practices are commendable, but only where patient-centered care remains central to working behaviors. It is important to remember that while this patient and examination may be one of a long list of examinations and patients for the radiographer during the day, for the patient it is likely to be their first, and the outcome of the X-ray could be life-changing depending on the clinical question and presenting symptoms. To address any tension between delivering services that are efficient and patient centered, radiographers and radiography managers should consider how resolving pathway bottlenecks and approaches to improve capacity might be introduced into the system to enable delivery of more patient-centered care. This could include embracing seven-day working, and the extension of appointment times during the day. Travel to, and parking at, hospitals is increasingly difficult, so appointments outside of typical working hours may be an attractive option for some patients.

Within the imaging department, patients may become frustrated where delays past the appointment time are experienced, particularly when not communicated. If there are delays to lists or long waiting times, it is important to keep patients and their carers updated, as many will worry about car parking, booked transportation, and failure to meet external responsibilities such as collecting children from school. There are many examples where this communication is done very effectively using noticeboards and live TV screens detailing estimated waiting times. However, these systems only work well when regularly updated, and so departmental staff needs to be clear on whose responsibility updates are, and the regularity of updates.

The delay in getting the results of the imaging examination or procedure can also be a challenge for patients. A patient may be very anxious about their results, especially if they believe they could be life-changing for both them and their families. Any suspicion of a potentially life-limiting condition will quickly raise anxieties over time to diagnosis and the next steps in the care pathway. This can lead to difficult conversations at the end of examination or procedure if the radiographer is not clear about the process for obtaining results and the time frame that the results will be available in. Remember, while the image acquisition event may be the start and finish of the examination for the radiographer, for the patient, the imaging pathway commenced with reporting the symptoms to the referring clinician and will only end with a diagnosis and treatment plan. Consequently, appreciating the patient's timeline will ensure radiographers understand areas for concern and the need for effective communication.

Time is a crucial aspect of delivering patient-centered care in imaging departments. This varies from the time pressures of a busy waiting room to the timeline through the imaging pathway, referral to results. It is essential that radiographers are mindful of the different impacts and perspectives of time by users of imaging services.

FUTURE PERSPECTIVE OF CARE

Going forward, as technology continues to develop, it is likely the radiographer's role will continue to evolve too. It is important that increased technological efficiency of imaging equipment does not dominate the radiographer's role and undermine the needs of the patient. However, with industry advertising new equipment as more cost efficient as a result of faster patient throughput, and a need for fewer radiographers, patient-centered care appears to be taking a back seat to efficiency savings. Whether this will translate into practice remains to be seen. There is an increasing awareness of the importance of patient choice and accountability, and radiographers need to ensure that choice is available within the imaging department. The use of care contracts which set out the patients' responsibilities and the healthcare professionals' responsibilities are becoming more common. Radiographers need to be aware of these contracts, and ensure that they are also signed up to them.

REFERENCES

Elliott R & Wright L (1999) Verbal communication: what do critical care nurses say to their unconscious or sedated patients? *Journal of Advanced Nursing* 29(6), 1412–1420.

Health Education England (2019) The Topol review. Available at: www.hee.nhs.uk/our-work/topol-review Accessed: 11.10.2019.

Health Education England, Skills for Health and Skill for Care (2017) Person centred approaches: Empowering people in their lives and communities to enable an upgrade in prevention, wellbeing, health, care and support. Available at: www.skills forhealth.org.uk/images/pdf/Person-Centred-Approaches-Framework.pdf Accessed: 11.10.2019.

Hello my name is (2020) A campaign for more compassionate care. Available at: https://www.hellomynameis.org.uk/ Accessed 19.03.2020

HM Government (2017) The Ionising Radiation (Medical Exposure) Regulations 2017. Available at: www.legislation.gov.uk/uksi/2017/1322/regulation/1/made Accessed: 11.10.2019.

Iacobuci G (2018) Medical model of care needs updating, say experts. *British Medical Journal* 360. doi: https://doiorg.ezproxy.derby.ac.uk/10.1136/bmj.k1034

Imison C, Curry N, Holder H, Castle-Clarke S, Nimmons D, Appleby J, Thorlby R & Lombardo S (2017) Shifting the balance of care: great expectations. Research report. Nuffield Trust. Available at: www.nuffieldtrust.org.uk/files/2017-02/shifting-the-balance-of-care-report-web-final.pdf Accessed: 11.10.2019.

NHS England (2014, October) Five Year forward view. Available at: www.england.nhs.uk/wp-content/uploads/2014/10/5yfv-web.pdf Accessed: 11.10.2019.

NHS England (2016) Sustainability and transformation partnerships. Available at: www.england.nhs.uk/integratedcare/stps/view-stps/ Accessed: 11.10.2019.

NHS England (2019, January 7) The NHS long term plan. Available at: www.longterm plan.nhs.uk/publication/nhs-long-term-plan/ Accessed: 11.10.2019.

Picker Institute Europe. Principles of person centred care. Available at: www.picker.org/about-us/picker-principles-of-person-centred-care/ Accessed: 11.10.2019.

Sackett D, Rosenberg W, Gray J, Haynes R & Richardson W (1996) Evidence based medicine: what it is and what it isn't. *British Medical Journal* 312, 71. doi: https//doi.org/10.1136/bmj.312.7023.71

The Health Foundation (2014) Patient-centred care made simple. Available at: www.health.org.uk/sites/health/files/PersonCentredCareMadeSimple.pdf Accessed: 11.10.2019.

Weaver R (2015) Reconciling evidence based medicine and patient centred care: Defining evidence based inputs to patient centred decisions. *Journal of Evaluation in Clinical Practice* 21(6), 1076–1080.

White K (2006) *The Sage Dictionary of Health & Society*. Sage Publications Limited. E-book. Canberra, Australia.

Williamson G, Jenkinson T & Proctor-Childs T (2009) *Nursing in Contemporary Healthcare Practice*. Learning Matters Ltd. E-book. Exeter, UK.

7 The Frequently Forgotten Pediatrics
Their Safety in the Clinical Setting

Chandra Makanjee

INTRODUCTION

The global emphasis on reductions in childhood mortality and meeting the sustainable developmental goals has resulted in significant gains in reducing childhood deaths around the world (Kassebaum et al., 2017). However, an epidemiologic shift has been noted, with a relative increase in deaths from injuries and a decline in deaths from poor nutrition and infections such as pneumonia and diarrheal diseases (Rivara, 2012; Kassebaum et al., 2017). Patient safety is a major priority for all healthcare professions and is a broadly conceptualized prevention of harm to patients. Radiography is an integral part of pediatric healthcare. It is well documented that this population group, apart from been vulnerable, is also highly susceptible to carconegic risks and have a longer life span to show manifestations (Emmanuel and Festus, 2018). Despite imaging modalities such as ultrasound (US), computed tomography (CT), and magnetic resonance (MRI) are frequently and increasingly utilized, general radiography remains an essential tool in the investigation of a wide spectrum of pediatric pathology, including neonates (Matthews et al., 2014). Recent studies have shown that though median radiation per patient is low, the range of radiation exposure is very wide. Some patients have high levels of radiation. These findings reinforce that this group is at risk for high radiation exposure and identifying these risk factors is important. The inherent dilemma is the characteristics of this vulnerable population group in determining radiation dose while sustaining image quality. In short, treating, managing, and optimizing pediatric care within general radiography remains pivotal to the pathway, which should not be underestimated (Matthews et al., 2014).

The importance of consistency and standardization in pediatric radiography, along with the preservation of image quality is important. Many of the prominent challenges to achieving this goal are technological and relate to the fact that most imaging equipment focuses on the imaging of adult patients, thus a requirement for extensive user modification in order to tailor these

systems for pediatric use. Adherence to protocols and guidelines cannot substitute the exercise of sound-based scientific judgment. This judgment in
ensuring consistency and standardization entails interrelated objective and
subjective factors by radiographers, which cannot be achieved without involving diverse role players within and out of the medical imaging team (Sánchez
et al., 2018). The aim of this chapter is to provide broad strokes on trends in
pediatric general radiography examination context.

THE REALITIES AND PITFALLS IN MAKING SCIENTIFICALLY SOUND EVIDENCE-BASED DECISIONS AND JUDGMENTS IN PEDIATRIC IMAGING

THE META TO MICRO ISSUES IN THE CONTINUUM OF CARE PROCESSES AND PROCEDURES

The theme of injury is chosen for this subsection to illustrate sound scientific
judgments at a microlevel in this population group. Kiragu et al. (2018) highlight
trauma as the leading cause of morbidity and mortality. Children in low- and
middle-income countries (LMICs) bear the greatest burden of unintentional and
intentional injury. The reality in the aforementioned countries is often the lack of
adequate resources to manage trauma. One of the avenues in managing trauma is
to follow trauma care protocols and to adapt treatments based on local resources.
At the meta level, it is pivotal to create awareness of injury prevention, regionaliz-
ing care and developing centers of excellence through multispecialty collaboration
within each country to improve outcomes and lowering trauma-related morbidity
and mortality globally. At the mesolevel is a commitment by governments in
LMICs in collaboration with international health organizations as well as partners in HICs to provide adequate healthcare services to their populations to be
a safeguard against the devastation of infectious diseases and lead to improved
outcomes for injured children (Kiragu et al., 2018). Further, the role of national
regulatory bodies to be relevant and filters to the micro level in conducting qual-
ity-assured imaging investigations without compromising the safety of the patient.
For example, very relevant and relating to injury and trauma in this population
group is that the 2008 RCR/RCPCH guidelines, which have been recently
updated and replaced by new 2017 RCR/SCoR SPA guidelines, recommending
an increased number of radiographic projections as part of the skeletal survey
(Hughes-Roberts et al., 2012; Hampel & Pascoal, 2012, 2018).

KNOWLEDGE GAP BRIDGING

With regard to safety there may be adequate knowledge of radiation protection in pediatric imaging among radiologists and radiographers but not so
with the referring pediatricians (Emmanuel & Festus, 2018). These authors
emphasize the need to bridge the knowledge gap by educating the pediatri-
cian, especially during residency training. Providers should strive to not

duplicate or order unnecessary tests whenever possible because these practices can result in undue radiation exposure and wasted resources. When placing a new request order, both the medical imaging team and referring health providers should be diligent in reviewing outstanding orders and results already returned to avoid any unnecessary radiation exposure. Thus, this is where evidence-based research plays a role. For example, studying the impact of electronic order entries on superfluous medical imaging within the trauma can capture unnecessary radiation usage to the pediatric patient.

THE INTERTWINED COMPLEXITIES AROUND PEDIATRIC DIAGNOSTIC IMAGING

Another issue is the role of general radiographs on pediatric trauma patients who undergo computed tomography (CT) may result in unnecessary radiation exposure. An example of this is where pelvic radiographs can be obtained after the CT interpretation, which would result in higher sensitivity of detecting fractures given the availability of the CT report at the time of radiographic interpretation. In other instances, CT and general radiographic examinations are obtained within one hour of each other. For example, the triage team may have initially ordered a general pelvis radiograph; however, after the trauma physician evaluation, a CT of the abdomen and pelvis is requested. It is important to note that while we discuss the role of diagnostic imaging in the urgent traumatic patient, less critically ill patients with lower suspicion for pelvic injury may benefit from nonemergent magnetic resonance imaging (MRI) evaluation on a case-by-case basis. This imaging workup may also be more appropriate for subacute trauma patients who present to the emergency department days after their injury. A limitation could be that MRI is not always available 24/7 and may result in unnecessary delays in diagnosis due to prolonged wait times, cost, and ability for patients to remain immobile for a period in order to acquire sound image quality (Alzen & Benz-Bohm, 2011). In short, appropriate triage and availability of imaging modalities becomes crucial.

Based on the above discussion, it becomes necessary to introduce didactic lectures on radiation protection and perhaps on radiobiology and appropriate choice of imaging modality for every clinical condition to medical school curriculum and the same extended to residency training in all aspects of clinical practice that requests radiological imaging. Assessment tools can be employed to determine the adequacy of the impact of such lectures on the pediatric residents.

THE ROLE OF RESEARCH

Research in pediatric diagnostic and interventional imaging has become extensive and expanded in recent year. As already addressed in the previous section, using evidence-based best principles and attaining high standards of safety while undertaking an imaging investigations remain paramount. Moreover, it leads to increased insights and better understandings in effective utilization of imaging examinations with minimal compromisation of the safety to the patient.

From a safety aspect, continuous research focuses on the cumulative effective dose and cancer risk associated with repetitive full spine imaging to young patients, as semiannual scans from low age onward are necessary for follow-up of scoliosis, for example. The EOS slot-scanning 2D/3D system (EOS Imaging, Paris, France) has the advantage of acquiring X-ray images using lowering radiation exposures than conventional digital systems and hence a reduction of stochastic cancer induction (a biplane X-ray imaging system, which allows the simultaneous acquisition of frontal and lateral images while the patient is in an upright weight bearing position). This effort to further decrease the radiation dose (μ-dose) has led to the development of protocols for EOS systems (Law et al., 2017).

Hampel and Pascoal (2018) undertook research to compared imaging techniques of each of the projections that compromises as part of a pediatric skeletal survey to establish dose reference levels (DRLs) for each projection for skeletal pediatric surveys.

COLLABORATIVE MULTIDISCIPLINARY ROLE PLAYERS IN ACHIEVING QUALITY OUTCOMES

Waddell and Connelly (2018) undertook a dose optimization study, which involved a project on cervical spine clearance in children. This study illustrated the impact of omitting radiographers from the education process of the project, which resulted in confusion and 'push back' in the beginning of the project. Additionally, there was also an issue of alignment in the quality of images, which was highlighted by radiographers. The surgical staff provided individual education to the radiographers when this situation occurred. The identified barriers include provider bias toward current practice, no current standardized protocol(s), communication among staff, equipment concerns, and techniques for capturing radiographs. Identifying possible barriers prior to implementation of the project was helped with brainstorming strategies to circumvent these issues as they presented. Continued multidisciplinary commitment, reinforcement of a cervical spine clearance protocol, and active engagement of staff lead to continued success in dose reduction in pediatric trauma patients. Involvement of stakeholders and providing education prior to a quality improvement are imperative for success. Findings of this study could make physicians as well as trauma and surgery teams conscious of the risk factors of radiologic studies. It is strongly suggested that coordination with radiology departments to ensure that pediatric protocols for imaging studies are in place to ensure the 'ALARA' principle is adhered (Kapileshwarkar et al., 2018).

THE IMPORTANCE OF PATIENT CENTRICITY

An imaging facility providing pediatric medical imaging services should ensure that dedicated protocols are developed and implemented for diagnostic imaging examinations. A study conducted in the Netherlands showed variation in radiation protection practices. Additionally, this is the rationale for standardizing and gently phasing fundamental steps to minimize radiation in pediatric patients. This includes delivery of optimum X-ray exposures (Hayre, 2016), accurate patient centering, and proper use of shielding where applicable. An integrated patient-centric approach avoids duplication of studies performed in other hospitals whereby central-imaging repositories can be used to reduce duplicate radiological studies but also could be used to monitor radiation exposures over time.

THE PEDIATRIC PATIENT CHARACTERISTICS

The proportions of a child's body differ considerably from those of an adult's body (Matthews et al., 2014). In general, an infant's body is shorter and broader than that of an adult. If the trunk of an infant or small child's body is X-rayed, the body shape inevitably means that larger areas of the body lie within the radiation field, or at least directly adjacent to it, and therefore more affected by scattered radiation. The peripheral skeleton generally has a relatively low radiosensitivity compared to the trunk, and generally results in lower effective doses than for body exposures. It is important to note that babies and younger children have highly radiosensitive red bone marrow in long bones, while older children and adults have red bone marrow mainly in the axial skeleton.

LOCATION OF AT-RISK TISSUE

There are also differences in the location of particularly at-risk tissues such as hematopoietic bone marrow. In adults, 74% (spine, ribs, pelvis) is located in the skeleton or the trunk, and only 9% in the extremities. In adults, 8% is located in the cranial bones, whereas in infants, it is 27%. In infants, 29% is located in the skeleton of the trunk and 35% in the extremities. This means that infants have large proportions of hematopoietic bone marrow in all parts of the body, including the extremities, hence an enhanced risk for children when compared to adults. It is generally accepted that children's tissues have a higher cell division rate, and thus cells are at an increased likelihood of stochastic cancer induction. An optimization strategy is required allowing for accurate diagnosis of subtle fractures, periosteal reactions, and early callus formation, while still conforming to the as low as reasonably achievable (ALARA) principle. Additionally, an essential difference between adult and pediatric medical imaging is the divergent disease profile of children and the resulting clinical questions regarding imaging (Alzen & Benz-Bohm, 2011).

THE MOST FREQUENT CHEST EXAMINATION

Premature neonates' survival has improved notably in the last decades due to the use of modern intensive care interventions. Such actions are usually accompanied by frequent radiographic examinations (Gois et al., 2019). Crealey et al.

(2018) report that 44% of neonatal radiographs mostly chest radiographs (77%) are performed out of hours. The rationale for chest radiographs for neonatal conditions can be due to respiratory distress, line position, respiratory deterioration, abdominal dissension, endotracheal tube position confirmation, cardiac assessment, ventilation assessment, bilious aspirator, pneumothorax/chest drain, fracture, sepsis assessment, and congenital abnormality.

OSSEOUS EVALUATIONS

Bone lesions are encountered in pediatric patients, with primary bone tumors representing the common neoplasm. Although cross-sectional imaging such as CT or MRI can be useful, general radiography continues to be a primary modality of choice for initial evaluation of osseous abnormalities (Vartevan et al., 2018).

ABDOMINAL GASTROINTESTINAL OBSTRUCTION (GIO)

The role of plain abdomen imaging among the pediatric population is highly dependent on the condition of the patient. The incidence of surgical emergency in neonates is between 1 and 4 per 100 births (Seth et al., 2015), and GIO is one of the most common surgical emergencies (Karami, 2008). In cases that are suspicious for upper GIO, contrast studies are necessary to enhance diagnostic accuracy, but in lower GIO, noncontrast radiographs are performed first because of its favorable accuracy for detection of the level of obstruction and if it is equivocal or nondiagnostic, then contrast agents are utilized (Afzal et al., 2019).

FOREIGN BODY INGESTION

Indication for foreign body (FB) ingestion is common in pediatric emergency services. It may be organic or inorganic, sharp or blunt, and usually passes through the digestive system. However, rarely may present with acute abdomen symptoms due to perforation or even acute appendicitis. It mostly occurs with sharp objects, whereas blunt objects might cause inflammation in intestinal segments with abscess, especially in the appendix. An abdominal radiograph has its value in localizing most radiopage FB objects (Ayaz et al., 2018). However, interpretation of the image may pose challenges. CT is useful in detecting the exact location but regarding inflammation and perforation ultrasonography may also not be helpful, which is best outlined by Ayaz et al. (2018).

INTENTIONAL AND UNINTENTIONAL INJURIES

With regard to the aforementioned is the mechanism of injury. Some injury patterns, according to Leonard et al. (2014), result in injury patterns and locations that are different from adult populus and could result in higher incidence of mortality (Leonard et al., 2014). Brown et al. (2018) refer to these as age-related injury patterns (Brown et al., 2018). The mechanism by which these injuries occur has a wider variation in children, as they are more vulnerable than adults to inflicted injury, sporting injury, and falls. The incidence of skateboard injuries is estimated to be 2–2.6% of all pediatric injuries and the rate of hospital admission of injured skateboarders is 2.9% (Ma et al., 2018). The most serious injuries to skateboarders involve head injuries and head injuries account for between 3% and 9% of all skateboard (ibid).

The World Health Organization (WHO) reports the top five etiologies for unintentional injuries due to road traffic accidents (RTAs): falls, burns, drowning, and poisoning (Chandran et al., 2010). Alarming numbers linked to injuries and deaths are also related to war, disasters, and child abuse. Although the prevalence and type of abuse may vary by location, the signs and symptoms are often similar. Children can present with symptoms such as retinal hemorrhage, subdural hemorrhage, and encephalopathy or evidence of diffuse axonal injury (Gonzalez & McCall, 2018). Additionally, child abuse could be considered when features are not consistent with the history provided, such as rib fractures, bruises of the ears, toes, or in the form of handprints (Kiragu et al., 2018). A helpful mnemonic for assessing which bruises are more concerning is the 'TEN 4' rule, which include Torso, Ear, Neck, and 4 Worrisome fractures involving posterior or lateral rib fractures, 'bucket handle' fractures, and fractures involving the sternum, spine, and scapula unless the child has been in a major motor vehicle or similar accident. In children, with elevated liver function tests, pancreatic enzymes, or otherwise unexplained hematuria, abuse involving abdominal trauma should be considered. Other signs and symptoms that should trigger a careful exam include unusual scars in the form of hand and belt prints, cigarette burns, burns in an unusual distribution, or swelling that is unexplainable such as swollen painful thigh in a young infant. Each of the seven strategies in the INSPIRE strategies, elucidated later, are important in prevention as well as specific treatment for the injuries sustained (Gonzalez & McCall, 2018). It is also important to differentiate abuse from traditional treatment and practices such as tattooing, cupping, and coining prevalent in many cultures (Lilley & Kundu, 2012; Kiragu et al., 2018). In suspected physical abuse (SPA), skeletal surveys are performed in order to detect bone fractures, which remain the most common clinical presentation of SPA, after soft tissue bruising and burns (Clarke et al., 2012). The current national guidelines presented by the Royal College of Radiologists and Society and College of Radiologists (RCR/SCoR) recommend a full skeletal survey with 25 projections for small children and 34 for large children. It is essential that the diagnosis of abuse is ascertained and not missed, as evidence shows that in 50–80% of fatal or near-fatal abuse cases, there was evidence of prior abusive injuries. Therefore, ensuring that images with optimal image quality for the diagnostic purposes are produced is essential (Hampel & Pascoal, 2018).

PEDIATRIC DIAGNOSTIC MEDICAL IMAGING EQUIPMENT AND CHARACTERISTICS

As discussed earlier, the pediatric age group is at greater risk to ionizing radiation due to the proliferation of cell division. Because these tissues are highly sensitive to radiation, poorly adapted radiological equipment to pediatric patients increases the effective dose of ionizing radiation.

EQUIPMENT TYPES

Crealey et al. (2018) comment on the use of portable computed radiography (CR) systems for neonates within the last decade affirming good working order with regular quality assurance in accordance with recommended guidelines. However, there was no integrated dose area product (DAP) meter and thus unable to record the individual radiation dose for each patient. Dose estimations were undertaken using 63 kVp (range: 52–66 kVp) and 2 mAs (range: 1.1–2.5) for a normal birth weight infant. These factors were adjusted with the size of the infant. Using these exposure factors as inputs, along with estimated beam geometry and technical factors from the X-ray system, patient doses were estimated using PCXMC VR 2.0.1.3, a personal computer X-ray Monte Carlo program. A phantom size of a zero age was selected, and the patient weight was adjusted to the median weights of the neonatal patients in UMHL. The weight selected for extreme low birth weight (ELBW) patients was 1 kilogram (kg), as this was the minimum weight accepted by compute program, PCXMC. The beam geometry was adjusted to represent appropriate field sizes for the clinical request and patient size and has been found useful in a similar study by Smans et al. (2008).

FILTRATION

The International Commission on Radiological Protection (ICRP) has recommended the use of additional filtration regarding pediatric imaging, excluding neonates and very small infants due to the low kVp used (ICRP, 2013). To avoid errors, additional beam filtration should be selected automatically when using the appropriate projection and age-group exposure preset (Knight, 2014). In a study by Al-Murshedi et al. (2019), they undertook a comparative analysis of radiation dose and low contrast detectability using routine pediatric chest radiography protocols. They identified that reasons for not adopting additional filtration was because general district hospitals did not specifically focus on the imaging of pediatrics. These authors also highlighted that identical incident air kerma (IAK) values on two different X-ray units, with different tube potentials and different levels of filtration, may generate differences in effective dose (Al-Mushedi et al., 2018). The fundamental reduction in X-ray flux emitted by the X-ray tube during pediatric imaging provides an opportunity for the medical physicist to recommend different X-ray tube voltages and added filtration, reduced pulse widths, or focal spot sizes that either improve image quality, reduce patient dose, or both. The medical physicist ensures that the desired acquisition parameters for pediatric imaging are incorporated into the configuration of the installed imaging device (Strauss et al., 2015).

AUTOMATIC EXPOSURE CONTROL

Automatic exposure control (AEC) is generally associated with higher radiation dose for pediatrics when compared with the manual exposure control. The likely reason for delivering higher radiation dose with the AEC for pediatric compared with that of the manual control may be because most AEC systems are not designed specifically for pediatric patients (Hintenlang et al., 2002). These systems have comparatively large and fixed ionization chambers and their size, shape, and location do not reflect the differences in body size in pediatric patients (ibid). A study on pediatric chest radiography reported the use of the AEC using all three chambers selected (both lateral and central) regardless of patient's age (Kostova-Lefterova, 2015). Some digital radiography (DR) units now contain a feature whereby AEC sensors inside of the collimated X-ray beam are activated to avoid underexposure.

ANTISCATTER GRIDS

Antiscatter grids improve image contrast at a cost of increased patient doses. The choice of utilizing a grid or not requires sound knowledge of grid characteristics, especially for pediatric digital imaging examinations. Antiscatter grids are generally not required for chest radiography of small children due to the small amount of scatter radiation. Recommendations, however, were made to use an antiscattering grid only for special indications in adolescents and never for children younger than 1 month. It is typically used for children who are able to cooperate and stand upright in front of the vertical stand (Kostova-Lefterova et al., 2015).

Beam attenuation by the table (for a constant detector exposure) will increase patient dose and/or reduce image contrast to noise ratio (CNR). This is due to the table adding another attenuating layer. This is more of an issue for tables of higher radio-opacity (such as cantilevered tables), and when using lower kVp parameters. Where possible, if wireless detectors are available or a wall detector, which can be flipped into a horizontal position, nongrid body exposures and distal extremity exposures taken through the table are avoided to optimize image quality.

IMAGE RECEPTOR SYSTEMS

The International Electrotechnical Commission (IEC) stated that the exposure index (EI) can be defined as a measure of a digital detector's response to radiation in the relevant image region (RIR) of an image acquired using DR (IEC, 2008). Generally, the EI is influenced by the patient size, artifacts, source to image receptor distance, collimation, centering, and imaging receptor size (Butler et al., 2010). The EI value simply indicates the level of exposure detected by the imaging receptor. Fore example, if a thick body part is imaged on a detector, the exposure technique must be adjusted upward to compensate for increased X-ray attenuation. If a thin patient is imaged on the same detector, the

X-ray technique must be reduced to achieve the same number of transmitted X-ray photons to the detector (same signal-to-noise ratio (SNR)), with the result being the same EI value, and at a lower dose to the patient. For a more detailed account of EI values in the clinical environment, please see Chapter 5.

All the digital detectors and computed radiography equipment should use multifrequency processing, which can maintain optimal image quality at lower mAs, and is thus useful for optimal pediatric digital imaging (Knight, 2014). For example, the Philips Digital Diagnost has default pediatric image processing algorithms to optimize image quality. Different algorithms are required for different age groups, grid status (in or out), and projection. Which may require custom image processing settings in liaison with the vendor's application specialist(s).

EXPOSURE TECHNIQUE CHART CONSIDERATIONS

According to Sánchez et al. (2017), default pediatric protocols are configured based on patient age. However, age does not adequately characterize the patients size, which remains a central determinant of proper imaging technique. The use of default age-based protocols could lead to suboptimal image quality, overexposure or underexposure, saturation of the digital detector, and nondiagnostic quality because of excessive noise (Hayre, 2016). Because the patient thickness determines the amount of radiation exposure that reaches the detector, without appropriately accounting for variations in thickness, it is impossible to achieve consistent image quality across patients for a given protocol.

AGE GROUPING AND SIZING

Knight (2014) modified size groups to be based around full-term baby to adult (large adolescent) sizes based on patient age. Depending on the type of imaging equipment, additional presets maybe required for each age group and projection to allow for age-group functionality. The following age groups were chosen, partially based on vendor recommendations and European DRLs: 0–6 months, 6–18 months, 18–36 months, 3–7 years (average 5 years), 8–12 years (average 10 years), 13–17 years (average 15 years), and 18 years. These age groups were based on exposures suitable for tissue thickness (in the direction of the X-ray beam) of a patient of 'average/standard size' in that age group for each projection. An overt challenge lies in the difficulty of comparing the pediatric intensive care unit (PICU) population with variability in weight, age, and body surface area (Kapileshwarkar et al., 2018). Further, Gois et al. (2019) used an infant's sex, birth weight, and time spent in the neonatal intensive care unit (NICU). The infant's weight was measured during hospitalization and plotted on a weight gain chart. A third-degree polynomial was used to determine the weight on the day of the radiographic examination. The examination type, tube voltage (kVp), current-time product (mAs), and focus-surface distance (FSD) of each examination were recorded. It is highly recommended that measurements, literature, and reviewing of imaging within each age group (to adjust exposures as necessary) be used for exposure chart development. Changes to the default radiographic parameters are required when patients are smaller or larger than average for their age group (Knight, 2014). Interestingly, Sánchez et al. (2017) used <1 day, 1–7 days, 1–4 weeks, 1–3 months, 3–6 months, 6–12 months, 12–24 months, 2–6 years, 6–10 years, and >10 years as a breakdown of functional age, which differs from Knight (2014). It is also important to use a standard measurement location for each projection (e.g., is a forearm measured at the elbow, mid-shaft, or wrist?).

USE OF CALIPERS

If using calipers, it could distress children and could be at risk of inappropriate use (e.g., measuring the chest thickness of adolescent girls). Considering increased examination times in busy clinics due to additional time taken to perform caliper measurements, infection control risks of using calipers with multiple patients and technical considerations with thickness-based exposure preset programming and linking with projection specific image processing remain paramount (Knight, 2014). Sánchez et al. (2017) used caliper-based thickness measurements to determine appropriate technique factors along with the EI value. Like other studies, their findings revealed a source of inconsistency in age-based protocols with significant overlap in size across age. Variations in EI are likely to result due to image segmentation errors due to patient positioning, supported with the effect of image-processing parameters or any metric of image quality apart from the EI (Sánchez et al., 2017).

However, according MacDoughal et al. (2018), calipers only measure a single point at any time and require physical contact. In its early uses, the device automatically measured body thickness and detected errors in positioning and motion before an X-ray was acquired. The software is designed to alert the radiographer when a potential problem arises, such as when the body part is not centered appropriately on the detector or whether the patient has moved off the AEC.

SOURCE-TO-IMAGE DISTANCE

It is well documented that increasing the source-to-image distant (SID) is as an effective method of reducing radiation dose to patients during medical imaging examinations. The effective dose reductions range from 6.7% by Monte Carlo calculations (Poletti & McLean, 2014) to a maximum of 44% by measurements on patients (Brennan & Nash, 1998). Increasing the SID from the traditional 100 to 130 cm results in reduction of 22% to 65%. This is highly dependent on demographic characteristics, the body part, and the desired projection undertaken examination, while maintaining sound image quality (Knight, 2014). For example, the entrance surface dose (ESD) for AP projections of the pelvis (31%), abdomen (25%), skull (23%), and spine radiographic (27%) examinations, respectively, was observed. For the lateral spine and skull projections, increasing the SID from 100 to 130 cm resulted in 22% and 35% reduction in ESD, respectively. In support of dose reduction, visual grading analysis (VGA) scores showed no statistically differences between the qualities of resultant images in both the 100 cm and 130 cm SIDs (Karami et al., 2017).

Despite a suggested increase in SID as a worthwhile dose-optimization tool clinically, a gap between the evidence and practice still exists; thus, it does not remain commonplace in many clinical settings. The traditional limitations discussed in the literature are related to equipment and radiographers' physical limitations. The physical dimensions of some radiography rooms may restrict increasing the SID in vertical and horizontal axis, for example, in a low-ceiling X-ray room. Moreover, there is an argument whereby frequently increasing the SID among radiography staff in the vertical axis may result in additional operational fatigue and ergonomic challenges, such as back pain. Furthermore, increased exposure output followed to increase of SID may result in a reduction in X-ray tube life; however, Brennan et al. (2004) affirm that such increases on tube loading is likely to have a negligible effect on the life of the X-ray tube.

The Concept of Shielding 'In Field' and 'Out of Field'

The area included within the primary X-ray beam is the area of interest and will be referred here as 'in-field'. The area excluded by the primary X-ray beam is referred to as 'out-of-field'. Accurate collimation reduces unnecessary radiation and improves image quality, which is important in the radiosensitive pediatric populus (Phelps et al., 2016; Tschauner et al., 2016). The gestation age (as a surrogate of weight and size) and the number of external lines or catheters (as a surrogate of disease severity) do not have any obvious effects on collimation. Even if poor collimation does not expose any additional parts of the body, the dose is still increased as a result of the higher doses proportion of scattered radiation, which can degrade image quality (Stollfuss et al., 2015; Khan et al., 2016). Unnecessary exposure of the abdomen, arms, or head during portable chest X-rays in the NICU is a relatively frequent phenomenon (Stollfuss et al., 2015). Emmanuel and Festus (2018) reported that during chest radiographs of neonates, the abdomen would often be included due to improper collimation.

In digital radiography, where the X-ray field collimation can be integrated in the examination protocol, if the default value is appropriately set, extreme cases of oversized collimation can occur, whereby the collimation automatically increases in size; thus, it remains paramount that the radiographer checks the field size upon selection of a protocol prior to exposure in order to ensure adequate collimation is still applied. In digital radiography, an additional pitfall is portrayed in the image processing and editing capabilities. For example, one issue is the ability to electronically mask, crop, or shutter radiographs post exposure (Hayre et al., 2019). It is generally accepted that improper use of cropping can not only hide diagnostically useful details but may also hide suboptimum collimation, thus a risk of concealing systematic overexposure of patients due to poor collimation practices (ibid). However, since the DAP value is a product of the dose and the field size (at any given distance from the tube focus), if a common DAP meter is used, it cannot be determined whether a given high DAP value is due to a high dose value or a large field size, or both. To differentiate between dose and field, the field size information in the digital imaging communications in medicine header can be used (Tsalafoutas, 2018). Suboptimal radiographic imaging practices regarding primary collimation can be concealed by electronic collimation and can result in systematic overexposures that may go unnoticed. Concerning the manually applied electronic collimation, whether for fully digital or hybrid units, it is mandatory to ensure that it is used with caution and only for masking the background noise and not areas of diagnostic interest, leaving the original radiation field edges discernible (Tsalafoutas, 2018).

The question of consideration is what is good practice in the application of lead shielding during projection radiography in pediatrics? Though shielding out-of-field body parts has been shown to provide only a small reduction in radiation dose, previous research with Monte Carlo simulations has shown how out-of-field shielding can increase skin entrance radiation dose (Matyagin et al., 2014). Often, efforts to reduce patient dose using gonadal shields increase dose due to incorrect placement resulting in a repeat X-ray exposures. With the use of the automatic exposure systems, photo timing cells may be covered, leading to increased radiation output of 63% for pediatrics (Kaplan et al., 2018). The recommendation for the use of specific gonad shielding on patients during medical diagnostic X-ray procedures is: When the gonads lie within or close to (about 5 cm from) the primary X-ray field despite proper beam limitation, the clinical objective of the examination should not be compromised. Lastly, when considering the use of contact shields to protect radiosensitive anatomy from unnecessary radiation exposure is the psychological benefit (McKenney et al., 2019). A contentious issue on radiation exposure during operative fixation of pediatric supracondylar fractures involving the humerus. In one study by Martus et al. (2018), the use of lead shielding was applied with measurements recorded at both thyroid and gonadal regions. The dose was then compared with patients who did not receive any shielding and this revealed no statistically significant differences (ibid). The equivalent dose to the thyroid and gonads was minimal and approximates daily background radiation. Shielding of radiosensitive organs is appropriate when practical to minimize cumulative lifetime radiation exposure, particularly in smaller patients and when longer fluoroscopy times are anticipated (Martus et al., 2018). Tissue weighting factors may be a stimulus for radiographers to reconsider which organs should be shielded. The sensitivity of different tissues permits consideration of its contribution to the effective dose, which is a single value figure determined by the risks of cancer, hereditary diseases, or genetic mutations due to ionizing radiation. Relative radiosensitivity is a concept implicit in tissue weighting factor (Wt): any factor greater than 0 indicates higher than average radiosensitivity, and the weighting factors of all the body tissues add up to 1.16 (Shanley & Matthews, 2018). However, evidence suggests that gonadal shielding is consistently employed incorrectly.

Wide Exposure Ranges in Pediatric Imaging

Kapileshwarkar et al.'s (2018) study included patients from the intensive care unit to identify risks of high radiation exposure. Despite their attempts to minimize radiation, almost 12% of the patients during a single hospitalization were exposed to above average annual background radiation. This average annual background radiation per person includes medical sources of radiation. Though median ionizing radiation per patient was low, for general X-ray examinations, a median interquartile range of 0.2 mSv and range of 0.01–29.4 mSv was identified. Though a wide range, these authors also report that daily routine chest radiographs may not influence length of stay or mortality; and does not contribute to risk of high radiation exposure. These authors also recommend direct dosimeters for a uniform approach for calculating radiation dose in patients. However, it is not feasible to quantify exposures from different types of studies with direct dosimeters. For instance, Gois et al. (2019) used values of incident air kerma (K_i), which were computed from the exposure parameters. The radiation outputs of the X-ray tube for distinct tube voltages (kVp) were registered using an ionization chamber (Model10 × 5–6, Radcal Corp.) with 6 cm^3 sensitive volume. The ionization chamber, attached to an electrometer, was placed 25 cm from the tabletop to restrain the effect of backscatter and at a distance of 100 cm from the X-ray focus. The exposure was collected, and the output was measured at 10 mAs in the tube potential range of 40–70 kVp, using 10 kVp increments. The Ki values for the kVp, mAs, and FSD used in examinations were computed with the equation with an overall uncertainty, better than 15%:

$$K_i = Y(\text{kVp}) \cdot \left(\frac{d}{FSD}\right)^2 \cdot \text{mAs}$$

Kapileshwarkar et al. (2018) suggest using web-based image repository such as Arkansas, which can serve a dual purpose of accessing previous studies as well as monitoring exposures. In this regard, Maempel et al. (2016) affirm that radiation exposures in the acute management of pediatric upper limb trauma is dependent on injury type and the procedure undertaken (though considerable variation can be expected for a given procedure). Higher radiation doses may be justified (in complex cases) and case review does not necessarily imply wrongdoing. Discussion of such cases could provide excellent learning opportunities (or identify deficiencies that can be addressed). Few studies have quantified the relative exposure of ionizing radiation using different spatial C-arm configurations. Findings reported DAP associated with a chest radiographs: 13 cGycm^2 for ages 1–5 years and 25 cGycm^2 for ages 6–10 years. Exposures for upper limb trauma were low with an exception of a high exposure 17.23 cGycm^2 following a supracondylar fracture, manipulation under anesthetic (MUA) and k-wire, equating to approximately one chest radiograph, which is known to be associated with negligible stochastic risk of inducing malignancy (1.3 per one million males and 1.9 per one million females).

Their plea is to minimize unnecessary risks for the legal/ethical requirements for audit. Protocols should include how the surgeon is informed intraoperatively if specific threshold exposures have been surpassed, and for mandatory review of each case surpassing the 95th percentile, to ascertain if the exposure was justified.

IMAGE QUALITY

Schaefer-Prokop et al. (2008) point out the interest in how well objective measures reflect the subjective grading of image quality and how much small differences in visual grading affect diagnostic performance under clinical conditions. The relationship between dose and image quality can be assessed quantitatively and qualitatively. Quantitative assessment involves objective physical measurements, such as modulation-transfer function, detective quantum efficiency (DQE) or contrast-to-noise ratio, and contrast-detail studies. Spatial resolution is important, particularly when trying to determine small, subtle fractures (Hampel & Pascoal, 2018).

Digital radiography systems perform differently in the detective quantum effectiveness; thus, imaging parameters used in one system may be directly transferable to another system. Image noise is defined as the standard deviation (SD) of the pixel value within a region of interest (Sun et al., 2012). Contrast-to-noise ration (CNR) is an important metric of image quality as it impacts on the visibility of subtle contrast variations and anatomical details (Hampel & Pascoal, 2018). Image noise is mainly determined by the mAs and is less dependent on the kVp changes, indicating that kVp can be reduced with reduction of radiation dose without compromising image quality (Sun et al., 2012). For example, prior to the optimization process, 52 kVp was used to image a structure only 1 cm thick, resulting in poor CNR. About 40 kVp can be used with no additional beam filtration up to 66 kVp for an adult size knee of 12 cm thickness. The EI value is then derived from the image SNR, which relates to absorbed energy at the detector (and not the patient) after an exposure event (Moore et al., 2012; Knight, 2014).

THE RELEVANCE OF AN OPTIMIZATION STUDY

Assessing current practice with national guidance and also exploring the potential to optimize tube voltage and X-ray beam filtration combined with improving image quality without detrimental impact on radiation dose to pediatric patients are important. Owing to the variety of X-ray units used clinically, X-ray examinations cannot be standardized. Therefore, optimization is necessary for each particular X-ray unit and for each X-ray examination (Sun et al., 2012). This is achieved through as already discussed in aforementioned sections like considering the type of equipment and its characteristics, for example, Ysio digital X-ray system, with a direct digital flat panel detector (Siemens Medical Solutions, Erlangen, Germany). This type of image receptor is composed of a cesium iodide scintillator coupled to a thin film transistor matrix with amorphous silicon technology. Note the pixel size, which is 144 μm, with an active detector area of 34 × 43 cm. The maximum spatial resolution of the system is mainly limited by its pixel size.

To simplify optimization strategies, exposure charts were developed around two different categories – body exposures and distal extremity exposures. Body exposures optimizations are used for projections involving the trunk, head, humerus, and femur. The trunk and head contain organs with relatively high radiosensitivity, including the lungs, colon, breast, gonads, stomach, thyroid, and eyes. An optimization strategy was required to simultaneously decrease patient dose and improve image quality (Knight, 2014) and then identify, retrospectively, the radiographic views with the highest dose (DAP): anteroposterior (AP) skull, lateral skull, lateral spine, AP abdomen. The AP abdomen projection could be selected for optimization because it includes part of the ribs, the spine, the pelvis and hips, all of which could show occult fractures unavailable in other views. Fractures in the spine can be easily overlooked due to the soft tissue overlay; hence, the AP abdomen projection must be of excellent image quality to enable small low contrast details to be observed. This may incur a relatively high radiation dose compared with the other projections (Hampel & Pascoal, 2018). By using a phantom of tissue equivalence and provide radiographic characteristics similar to those offered by human tissue one can establish the imaging parameters that are optimal without compromising the CNR, which can then be validate in the clinical setting.

ETHICAL DILEMMAS AROUND IMMOBILIZATION AND RESTRAINING

According to Article 12 from the United Nations Convention on the Rights of the Child, children have the right to express their opinion and be heard regarding decisions involving their own wellbeing (United Nations. Commission on Human Rights, 1991). A child refusing to be restrained has the right to be heard and respected. Furthermore Article 3 from the Convention on the Rights of the Child and the Australian Human Rights Commissions Act (1986) highlights that it is the radiographer's responsibility to act in the best interest of a child. In a radiographic procedure, the best interest should be considered while respecting their voice and ability to express their wishes. This will allow for appropriate consent to be gained, thereby assisting in creating a clearer distinction between 'immobilization' and 'restraint' in pediatric imaging. Ng and Doyle's (2018) literature review delved into aspects of immobilization by defining immobilization as 'rendering a child incapable of moving with the child's consent'. Within medical imaging, immobilization is used to keep a child still during an examination to avoid increased radiation dose to the child due to motion artifact and, therefore, repeated imaging. Since the use of force can harm patients physically through skin tears and emotionally through anxiety and fear of future radiographic examinations, it is suggested that using restraint remains a last resort and for only 5–10 minutes at a time (ibid).

Restraining children still within general imaging is frequently performed using Velcro straps, distraction techniques, sandbags, and the support of parents/carers (Noonan et al., 2017). Across three studies published in Australia, Kuwait, and the United Kingdom (UK) parental holding remains the most common technique for ensuring the child remains still (Ashti et al., 2012). Parental holding is advantageous due to the parent's ability to influence the emotions of the child, thus able to reduce the child's anxiety. However, in examinations where the parent is more anxious than the child, this may prove counterproductive. Distraction techniques using toys, blankets, and verbal communication are commonly used in Australia for 2–5 year-old children. Compared with other restraint techniques, distraction does not require the use of physical force, thereby falling into the 'immobilization' category. This proves effective for avoiding physical injury to the child and emotional trauma during future radiographic examinations. However, in general X-ray, parental or staff holding is still seen to be preferred over distraction (Noonan et al., 2017). This alludes to an explanation that distraction is only effective to a certain degree whereby children remain adamant on not undergoing an examination. As the Pigg-O-Stat completely encapsulates the child, this force can amount to 'restraint', whereas the Fuller Chair uses Velcro strapping and only imparts force on the necessary joints thereby falling under 'immobilization'. Thus, balancing the amount of force to avoid 'restraint' justifies the use of such devices, but may often be difficult to achieve. The inability to control the amount of force in commercial immobilization devices could be a reason for its minimal use clinically (Linder, 2017). However, the process of gaining consent for immobilization may not be an easy task, as radiographers are required to judge the child's competency and the child's ability to understand the given information. Studies highlight that the use of restraint decreases with age due to the child's ability to comprehend examination instructions. At an age where pediatric patients make their own informed decisions, they are termed 'Gillick competent' and has been adopted into Australian Law by the High Court of Australia with Victorian legislation stating that a child can refuse restraint if the medical practitioner/radiographer deems the child competent. This is an issue as the assessment of Gillick competency can vary based on the radiographer's ethical beliefs and morals (Noonan et al., 2017). Moreover, a radiographer's ability to accurately discern competency from incompetency in pediatrics remains unknown.

ISSUES AROUND EFFECTIVE COMMUNICATION

A recent article by Bibbo (2018) delves into inconsistencies on conveying radiation risk to the patients and/or carers. So, each communicator may convey different information about the level of risk for the same radiation procedure, leaving the consumer confused and frustrated. Communication of risk is inherently challenging. Any miscommunication could result in misunderstandings, mistrust, and also a tendency to overestimate small risks and underestimate large and serious risks (Bibbo, 2018). For effective communication, it is important that the risk communicator has a good knowledge of radiation and its deleterious effects, and the most appropriate method to convey the risk information. The information should be conveyed in a clear, unbiased, open, and consistent way without using jargon. The communicator should understand how people think and respond to risks, particularly since the patient/carer may be under stress because of the patient's medical state, and under stress if patient/carer has difficulties hearing and/or understanding what the communicator is trying to convey. The benefit derived from the imaging procedure should always be emphasized.

EDUCATION

From an educational perspective, a majority of programs have variations mainly in the conduct of clinical practice. Some healthcare organizations, such as those in Singapore, have separate children and adult hospitals, which means that some radiographers may have limited exposure to pediatric examinations. That reality can cause an impact on the knowledge, skills, and competencies acquired by students. For example, pediatric radiography was not an option for 24% of Norwegian students (Dos Reis et al., 2018). Further, it is also highlighted that approximately 50% of radiology practitioners and radiographers had undertaken a maximum of 20 hours of radiation protection education and training, which is regarded inferior to the recommended 30–50 hours for radiology practitioners and 100–140 hours for radiographers (ICRP, 2009). It does not reflect the number of hours most radiographers perform as part of their undergraduate radiography course program. For this reason, the author is of the opinion that participants may have underreported the number of hours of radiation protection education and training received, possibly by overlooking several topics that are interrelated to physiological/pathological processes and or radiology/radiography principles (Portelli et al., 2018). Other intrinsic factors, such as educational level and awareness of the radiographer, may be responsible for the large variation documented. The intrinsic factors, such as the education and awareness of the radiographer performing X-ray examinations in ICUs, may also be relevant (Stollfuss et al., 2015). From a clinical perspective, the number of optimal images for each individual radiographer may vary significantly and may not correlate with the years of experience in performing X-rays. However, while these factors are difficult to measure on an objective basis, the recommendation is an individualized education and training for radiographers, to improve image quality and limit ionization radiation in pediatric radiography where applicable (Stollfuss et al., 2015).

CONCLUSION

This chapter provides a holistic account for pediatric general radiography and its dimensions in the modern general diagnostic imaging contexts. It has discussed current trends such as the implementation of quality of improved projects and the use of an evidence-based research approach. The evolved method of compiling exposure technique charts for pediatrics with use of optimization studies illustrates the importance of research-informed practices. This approach supports the multifaceted approach to minimizing radiation dose to pediatric patients transnationally. Furthermore, the findings support previously established calls on radiographers to measure patient depth in order to objectively guide pediatric exposures. Results show that diagnostic image quality does not necessarily have to be optimal image quality, and this highlights that the radiographer's judgement remains an important contribution toward optimization of practices. Though the achievement of diagnostic images in pediatric radiography is highly dependent on a radiographer's skill level, it is equally important that multidisciplinary team members take on roles and responsibilities concerning dose optimization approaches for this vulnerable group.

REFERENCES

Afzal, N., Malek, A. and Abasi, N. (2019). Comparison of abdominal x-ray findings and results of surgery in neonates with gastrointestinal obstruction. *International Journal of Pediatrics*, 7(1), pp. 8877–8880. DOI: 10.22038/ijp.2018.34584.3042.

Al-Murshedi, S., Hogg, P., Meijer, A., Erenstein, H. and England, A. (2019). Comparative analysis of radiation dose and low contrast detail detectability using routine paediatric chest radiography protocols. *European Journal of Radiology*, 113, pp. 198–203.

Alzen, G. and Benz-Bohm, G. (2011). Radiation protection in pediatric radiology. *Deutsches Arzteblatt International*, 108(24), pp. 407–414. doi:10.3238/arztebl.2011.0407.

Ashti, M., Al-Abbad, M., Faleh, A. and Al-Ostath, S. (2012). Current immobilization implementation of pediatric patients in five major public hospitals in Kuwait: a prospective study into policies and guidelines for radiology departments. *Indian Journal of Innovations and Developments*, 1(8), pp. 647–652.

Ayaz, M., Aslan, A., Gercel, G., Özkanlı, S.Ş. and Durakbaşa, Ç.U. (2018). A pediatric foreign body appendicitis can cause a pitfall in imaging. *Journal of Diagnostic Medical Sonography*, 34(4), pp. 299–302.

Bibbo, G. (2018). Effective doses and standardised risk factors from paediatric diagnostic medical radiation exposures: information for radiation risk communication. *Journal of Medical Imaging and Radiation Oncology*, 62(1), pp. 43–50.

Brennan, P. and Nash, M. (1998). Increasing FFD: an effective dose-reducing tool for lateral lumbar spine investigations. *Radiography*, 4(4), pp. 251–259.

Brown, P., Munigangaiah, S., Davidson, N., Bruce, C. and Trivedi, J. (2018). A review of paediatric cervical spinal trauma. *Orthopaedics and Trauma*, 32(5), pp. 288–292.

Butler, M.L., Rainford, L., Last, J. and Brennan, P.C. (2010). Are exposure index values consistent in clinical practice? A multi-manufacturer investigation. *Radiation Protection Dosimetry*, 139(1–3), pp. 371–374.

Chandran, A., Hyder, A.A. and Peek-Asa, C. (2010). The global burden of unintentional injuries and an agenda for progress. *Epidemiologic Reviews*, 32(1), pp. 110–120.

Cherian, T., Mulholland, E.K., Carlin, J.B., Ostensen, H., Amin, R., Campo, M.D., Greenberg, D., Lagos, R., Lucero, M., Madhi, S.A. and O'Brien, K.L. (2005). Standardized interpretation of paediatric chest radiographs for the diagnosis of pneumonia in epidemiological studies. *Bulletin of the World Health Organization*, 83, pp. 353–359.

Clarke, N.M., Shelton, F.R., Taylor, C.C., Khan, T. and Needhirajan, S. (2012). The incidence of fractures in children under the age of 24 months–in relation to non-accidental injury. *Injury*, 43(6), pp. 762–765.

Crealey, M.R., Bowden, L., Ludusan, E., Pathan, M., Kenny, B., Hannigan, A. and Philip, R.K. (2018). Utilization of conventional radiography in a regional neonatal intensive care unit in Ireland. *The Journal of Maternal-Fetal & Neonatal Medicine*, pp. 1–7.

Dave, J.K., Jones, A.K., Fisher, R., Hulme, K., Rill, L., Zamora, D., Woodward, A., Brady, S., MacDougall, R.D., Goldman, L. and Lang, S. (2018). Current state of practice regarding digital radiography exposure indicators and deviation indices: report of AAPM Imaging Physics Committee Task Group 232. *Medical Physics*, 45 (11), pp. e1146–e1160.

Dos Reis, C.S., Pires-Jorge, J.A., York, H., Flaction, L., Johansen, S. and Maehle, S. (2018). Curricula, attributes and clinical experiences of radiography programs in four European educational institutions. *Radiography*, 24(3), pp. e61–e68.

Emmanuel, I.O. and Festus, E.O. (2018). Radiation protection: what do the Nigerian paediatric residents know? *International Journal of Tropical Disease & Health*, pp. 1–8.

Gois, M.L.C., Schelin, H.R., Denyak, V., Bunick, A.P., Ledesma, J.A. and Paschuk, S.A. (2019). Human factor in exposure from conventional radiographic examinations in very and extremely low birth weight patients. *Radiation Physics and Chemistry*, 155, pp. 31–37.

Gonzalez, D. and McCall, J.D. (2018). Child abuse and neglect. In: *StatPearls [Internet]*. Stat Pearls Publishing.

Hampel, J.R. and Pascoal, A. (2018). Comparison and optimization of imaging techniques in suspected physical abuse paediatric radiography. *The British Journal of Radiology*, 91(1083), p. 20170650.

Hayre, C.M. (2016). 'Cranking up', 'whacking up' and 'bumping up': x-ray exposures in contemporary practices. *Radiography*, 22(2), pp. 194–198.

Hayre, C.M., Blackman, S., Eyden, A. and Carlton, K. (2019). The use of digital side markers (DSM) and cropping in digital radiography. *Journal of Medical Imaging and Radiation Sciences*, 50(2), pp. 234–242.

Hintenlang, K.M., Williams, J.L. and Hintenlang, D.E. (2002). A survey of radiation dose associated with pediatric plain-film chest x-ray examinations. *Pediatric Radiology*, 32(11), pp. 771–777.

Hughes-Roberts, Y., Arthurs, O.J., Moss, H. and Set, P.A.K. (2012). Post-mortem skeletal surveys in suspected non-accidental injury. *Clinical Radiology*, 67(9), pp. 868–876.

ICRP. (2009). Publication 113: education and training in radiological protection for diagnostic and interventional procedures. *Ann ICRP*, 39(5).

ICRP. (2013). Radiological protection in paediatric diagnostic and interventional radiology. ICRP Publication 121. *Ann ICRP*, 42, pp. 1–63.

International Electrotechnical Commission. (2008). *Medical electrical equipment – exposure index of digital x-ray imaging systems-part 1: definitions and requirements for general radiography*. Geneva, Switzerland: IEC, pp. 62494–1.

Jamil, A., Mohd, M.I. and Zain, N.M. (2018). The consistency of exposure indicator values in digital radiography systems. *Radiation Protection Dosimetry*, 182(4), pp. 413–418.

Kapileshwarkar, Y.S., Smith, L.T., Szpunar, S.M. and Anne, P. (2018). Radiation exposure in pediatric intensive care unit patients: how much is too much? *Clinical Pediatrics*, 57(12), pp. 1391–1397.

Kaplan, S.L., Magill, D., Felice, M.A., Xiao, R., Ali, S. and Zhu, X. (2018). Female gonadal shielding with automatic exposure control increases radiation risks. *Pediatric Radiology*, 48, pp. 227–234.

Karami, V., Zabihzadeh, M., Danyaei, A. and Shams, N. (2016). Efficacy of increasing focus to film distance (FFD) for patient's dose and image quality in pediatric chest radiography. *International Journal of Pediatrics*, 4(9), pp. 3421–3429. doi:10.22038/ijp.2016.7319.

Karami, V., Zabihzadeh, M., Shams, N. and Gilavand, A. (2017). Optimization of radiological protection in pediatric patients undergoing common conventional radiological procedures: effectiveness of increasing the film to focus distance (FFD). *International Journal of Pediatrics*, 5(4), pp. 4771–4782.

Kassebaum, N.J., Kyu, H.H., Zoeckler, L., Olsen, H.E., Thomas, K., Pinho, C., et al. (2017). Child and adolescent health from 1990 to 2015 findings from the global burden of diseases, injuries, and risk factors 2015 study. *JAMA Pediatr*, 171, pp. 573–592. DOI: 10.1001/jamapediatrics.2017.0250.

Khan, A.M., Shera, T.A., Choh, N.A., Wani, G.M. and Ahmad, Z. (2016). Radiation protection in pediatric radiology. *JMS Skims*, 19(1), pp. 39–40.

Kiragu, A., Mwarumba, N., Adesina, A., Gidado, S., Dunlop, S., Mwachiro, M., Gbadero, D. and Slusher, T. (2018). Pediatric trauma care in low resource settings: challenges, opportunities, and solutions. *Frontiers in Pediatrics*, 6, p. 155.

Knight, S.P. (2014). A paediatric X-ray exposure chart. *Journal of Medical Radiation Sciences*, 61(3), pp. 191–201.

Kostova-Lefterova, D., Taseva, D., Hristova-Popova, J. and Vassileva, J. (2015). Optimisation of paediatric chest radiography. *Radiation Protection Dosimetry*, 165(1–4), pp. 231–234.

Ladia, A.P., Skiadopoulos, S.G., Karahaliou, A.N., Messaris, G.A., Delis, H.B. and Panayiotakis, G.S. (2016). The effect of increased body mass index on patient dose in paediatric radiography. *European Journal of Radiology*, 85(10), pp. 1689–1694.

Law, M., Ma, W.K., Chan, E., Mui, C., Ma, V., Ho, W.Y., Yip, L. and Lam, W. (2017). Cumulative effective dose and cancer risk of pediatric population in repetitive whole-body scan using dual-energy x-ray absorptiometry. *Journal of Clinical Densitometry*.

Lilley, E. and Kundu, R.V. (2012). Dermatoses secondary to Asian cultural practices. *International Journal of Dermatology*, 51, pp. 372–379. doi:10.1111/j.1365-4632.2011.05170.x.

Linder, J.M.B. (2017). Safety considerations in immobilizing pediatric clients for radiographic procedures. *Journal of Radiology Nursing*, 36(1), pp. 55–58.

Ma, N., Mills, S., McBride, C., Kimble, R. and Redmond, M. (2018). Neurological injuries from skateboards in paediatric and adolescent populations: injury types and severity. *ANZ Journal of Surgery*, 88(4), pp. 337–340.

MacDougall, R.D., Scherrer, B. and Don, S. (2018). Development of a tool to aid the radiologic technologist using augmented reality and computer vision. *Pediatric Radiology*, 48(1), pp. 141–145.

Maempel, J.F., Stone, O.D. and Murray, A.W. (2016). Quantification of radiation exposure in the operating theatre during management of common fractures of the upper extremity in children. *The Annals of the Royal College of Surgeons of England*, 98 (7), pp. 483–487.

Martus, J.E., Hilmes, M.A., Grice, J.V., Stutz, C.M., Schoenecker, J.G., Lovejoy, S.A. and Mencio, G.A. (2018). Radiation exposure during operative fixation of pediatric supracondylar humerus fractures: is lead shielding necessary? *Journal of Pediatric Orthopaedics*, 38(5), pp. 249–253.

Matthews, K., Brennan, P.C. and McEntee, M.F. (2014). An evaluation of paediatric projection radiography in Ireland. *Radiography*, 20(3), pp. 189–194.

Matyagin, Y., Collins, P., Ruwoldt, S., Chew, S. and West, J. (2014). Effectiveness of gonad shields: a Monte Carlo evaluation. Royal Australian and NewZealand College of Radiologists Combined Science Meeting.

McKenney, S., Gingold, E. and Zaidi, H. (2019). Gonadal shielding should be discontinued for most diagnostic imaging exams. *Medical Physics*, 46(3), pp. 1111–1114.

Misiura, A.K., Nanassy, A.D. and Urbine, J. (2018). Usefulness of pelvic radiographs in the initial trauma evaluation with concurrent CT: is additional radiation exposure necessary? *International Journal of Pediatrics*.

Moloney, F., Fama, D., Twomey, M., et al. (2016). Cumulative radiation exposure from diagnostic imaging in intensive care unit patients. *World J Radiol*, 8, pp. 419–427.

Moore, Q.T., Don, S., Goske, M.J., Strauss, K.J., Cohen, M., Herrmann, T., MacDougall, R., Noble, L., Morrison, G., John, S.D. and Lehman, L. (2012). Image gently: using exposure indicators to improve pediatric digital radiography. *Radiologic Technology*, 84(1), pp. 93–99.

Ng, J.H.S. and Doyle, E. (2018). Keeping children still in medical imaging examinations-immobilisation or restraint: a literature review. *Journal of Medical Imaging and Radiation Sciences*.

Noonan, S., Spuur, K. and Nielsen, S. (2017). Immobilisation in Australian paediatric medical imaging: a pilot study. *Radiography*, 23(2), pp. e34–e40.

Phelps, A.S., Gould, R.G., Courtier, J.L., Marcovici, P.A., Salani, C. and MacKenzie, J. D. (2016). How much does lead shielding during fluoroscopy reduce radiation dose to out-of-field body parts? *Journal of Medical Imaging and Radiation Sciences*, 47(2), pp. 171–177.

Poletti, J. and McLean, D. (2014). The effect of source to image-receptor distance on effective dose for some common x-ray projections. *The British Journal of Radiology*, 78(2005), pp. 810–815.

Portelli, J.L., McNulty, J.P., Bezzina, P. and Rainford, L. (2016). Paediatric imaging radiation dose awareness and use of referral guidelines amongst radiology practitioners and radiographers. *Insights into Imaging*, 7(1), pp. 145–153.

Rivara, F.P. (2012). Prevention of death and disability from injuries to children and adolescents. *International Journal of Injury Control and Safety Promotion*, 19, pp. 226–230. doi:10.1080/17457300.2012.686919.

Sánchez, A.A., Reiser, I., Baxter, T., Zhang, Y., Finkle, J.H., Lu, Z.F. and Feinstein, K. A. (2018). Portable abdomen radiography: moving to thickness-based protocols. *Pediatric Radiology*, 48(2), pp. 210–215.

Schaefer-Prokop, C., Neitzel, U., Venema, H.W., Uffmann, M. and Prokop, M. (2008). Digital chest radiography: an update on modern technology, dose containment and control of image quality. *European Radiology*, 18(9), pp. 1818–1830.

Seth, A., Chanchlani, R. and Rakhonde, A.K. (2015). Neonatal gastrointestinal emergencies in a tertiary care centre in Bhopal, India: A prospective study. *IJSS*, 1(2), Wyllie R. 'Intestinal atresia, stenosis and malrotation', In: Kliegman RM, Behrman. 2. Karami, H, Alamsahebpour, A, Ghasemi, M, Khademloo M. Diagnostic value of barium enema in hirschsprung s disease. JBUMS. 2008; 10 (1): 54–59.

Shanley, C. and Matthews, K. (2018). A questionnaire study of radiography educator opinions about patient lead shielding during digital projection radiography. *Radiography*, 24(4), pp. 328–333.

Smans, K., Struelens, L., Smet, M., et al. (2008). Patient dose in neonatal units. *Radiation Protection Dosimetry*, 131(1), pp. 143–147.

Society and College of Radiographers, Royal College of Radiologists. (2017). *The radiological investigation of suspected physical abuse in children guidance.* London: The Royal College of Radiologists.

Srinivasa, S. and Patel, S. (2018). A study on distribution pattern of lower respiratory tract infections in children under 5 years in a tertiary care centre. *International Journal of Contemporary Pediatrics*, 5(2), p. 456.

Stollfuss, J., Schneider, K. and Krüger-Stollfuss, I. (2015). A comparative study of collimation in bedside chest radiography for preterm infants in two teaching hospitals. *European Journal of Radiology Open*, 2, pp. 118–122.

Strauss, K.J., Frush, D.P. and Goske, M.J. (2015). Image gently campaign: making a world of difference. *Medical Physics*, 3(2).

Sun, Z., Lin, C., Tyan, Y. and Ng, K.H. (2012). Optimization of chest radiographic imaging parameters: a comparison of image quality and entrance skin dose for digital chest radiography systems. *Clinical Imaging*, 36(4), pp. 279–286.

Tapiovaara, M. (2008). *PCMXC- A Monte Carlo Programme for calculating patient doses in medical x-ray examinations.* 2nd ed. Volume STUK-A 231. Helsinki: Sateilyturvakeskus.

Tsalafoutas, I.A. (2018). Electronic collimation of radiographic images: does it comprise an overexposure risk? *The British Journal of Radiology*, 91(1086), p. 20170958.

United Nations. Commission on Human Rights (1991). *Convention on the rights of the child* (Vol. 64, No. 11). United Nations Publications.

Vartevan, A., May, C. and Barnes, C.E. (2018). Pediatric bone imaging: differentiating benign lesions from malignant. *Applied Radiology*, 47(7), pp. 8–15.

Waddell, V.A. and Connelly, S. (2018). Decreasing radiation exposure in pediatric trauma related to cervical spine clearance: a quality improvement project. *Journal of Trauma Nursing*, 25(1), pp. 38–44.

8 Patient Care
'It's Just not for Me, It is for Everybody'

Chandra Makanjee, Deon Xu and Drishti Sarswat

INTRODUCTION

According to Undeland and Malterud (2002, p. 145), the term 'diagnosis' is derived from a Greek word meaning 'decision'. To diagnose is to make a decision. Patients (PTs) are often referred for a diagnostic imaging by medical practitioners in order to seek an opinion, which answers the clinical question. One cannot ignore how healthcare providers interact between the PT directly and/or indirectly and how this shapes the PT's expectations and interactional processes that are inherently interrelated and interdependent on text and technology. Its actual value lies in the continuum of care context from the point of referral to the outcomes of the investigation embedded within the clinical pathways as part of their health outcomes and integrated with the quality of life expectations and experiences. The added value in providing meaningful care goes much wider, taking into consideration the patient as a person. That is, from a lifeworld perspective of importance is the family and the society (Schout and de Jong, 2018). It is argued that sustaining a sound radiographic practice by applying best practice principles with speed in obtaining an optimum image does not necessarily translate into a quality-based valued outcome. The impact of the diagnostic imaging encounter must be considered within the ambit of patient-centric care approach where patient preferences and values need to be respected throughout medical encounter, of which medical imaging remains an integral part (McGovern, et al., 2018; Pahade, et al., 2018).

Because of the complex nature of the imaging encounter, the structure proposed is to provide an overview of the quality aspects regarding a diagnostic medical imaging encounter, which informs the backdrop. Followed by the referral of an imaging investigation, an important juncture is to enable a sound radiographic investigation. This chapter is based on a research projects which entailed shadowed case studies and a recent qualitative study to illustrate that the situational context within which the imaging encounter occurs. In short, patient perspectives on undergoing medical imaging are a temporal encounter. Their expectations and experiences are shaped and reshaped not only by meeting a diverse range of professionals from a biomedical scientific lifeworld, but they also deal with the unfamiliar technological world, that is, in this instance, the diagnostic imaging encounter that gives them hope leading to the psychosocial angle that could have a negative or positive influence on their quality of life with a desire to return to normalcy (Lovric and Makanjee, 2018).

It is important to capture the PT pathway from his or her entry into the healthcare system (i.e., even before consulting the radiographic practitioner) up to the point of diagnosis in order to present a holistic account of referral patterns, interactions, and decision-making processes within the continuum care processes and procedures. Furthermore, it indicates how these actions influences both PT and healthcare professionals' expectations, experiences, meanings, and understandings of imaging processes and procedures in relation to ensure a quality health outcome.

This chapter is based on the PT voices of their lived expectations and experiences of general radiographic investigations. How they construct their meaning and understandings of general radiography investigations is part of an episodic medical encounter in two separate countries. The studies received ethical approvals from various institutes for dissemination of findings. The findings will be used together with literature support to illustrate the quality of the PT care in an endeavor to further sustain sound radiographic imaging.

SETTING THE SCENE FOR MANAGING QUALITY WITHIN THE PT CARE

Sustaining quality and safety of healthcare remains an international challenge (Hogden, et al., 2018). Healthcare can often require alteration of processes in a complex social system over time in both predictable and unpredictable ways (Berwick, 1998). Within healthcare a diverse aspect of needs has to be considered in managing quality of services. The quality framework must be realistic, consistent, and manageable, aligning with the general mission and value structure. Therefore, monitoring the consistency of quality integrates into a wider institutional process. To enable a quality outcome entails accounting for the PT encounter at each point of contact. The complexity comes when every stakeholder has views on what constitutes a quality outcome, which alert to the importance of aligning and defining quality outcomes. Taylor, McNicholas, Nicolay, et al. (2013) are of the view that effective interventions are complex and multifaceted, therefore, an iterative process to adapt to the local context. Finding effective quality improvement methods is essential for delivery of high-quality valued care. This process is iterative and needs to be tested and evaluated. Central to the value structure of any institution or facility is the delivery of quality goods and/or services (Abdous, 2009). The assumption is that these processes are embedded within the daily operations of an institution or facility to deliver quality services (Lagrosen, et al., 2004).

Quality management (QM) within the diagnostic medical imaging context entails many different aspects and may be defined in many ways and forms different approaches. The QM framework is likely to include processes relating to quality assurance (QA) and quality control (QC) of imaging equipment (Papp, 2015). QA is a formalized management structure used to identify gaps and shortcomings in the aforementioned processes (Harvey, et al., 2016). QA is used to ensure the safe, effective, efficient, and timely provision of quality professional service and PT care (Zygmont, et al., 2017).

The challenge, and an essential aspect in making the transition from these abstract ideals of PT-focused 'quality' to the implementation of in-situ QA frameworks, is to embed it within a QM process. According to Lagrosen, et al. (2004), it is vital to contextualize the meaning of QA within the specific situation it is being applied to. This transition requires carefully coordinated and aligned plans to ensure that quality is sustained in terms of institutional structure, operational process, and procedural outcomes, which includes the demand for equitable access to imaging services (Harvey, et al., 2016; Ngoya, et al., 2016). The QC, of a medical imaging facility, entails QC of the equipment and its accessories, image quality, and PT safety aspects. So, a timely access to and delivery of integrated and appropriate imaging procedures in a safe and responsive facility and prompt delivery, followed with timely radiological reports, remain paramount (Harvey, et al., 2016; Zygmont, et al., 2017). For PTs, to receive the correct imaging projections, in the correct order, performed with precision and reported in the right format, can ultimately lead to sound clinical decision-making, irrespective of income, physical location, and cultural background. Furthermore, it is important to render uninterrupted and coordinated care across facilities and practitioners. For instance, the availability of and access to relevant clinical history, indications and findings of previous medical imaging studies or interventions, and the opportunity to discuss this with the referring practitioner and their PT is an essential component, which can significantly influence the imaging study.

Within the general radiography context, it is the radiographer and/or the radiologist together with the referrer who select the most appropriate imaging investigation for the clinical condition, based on evidence-based practice guidelines, while taking into consideration the avoidance or minimization of actual or potential harm from medical imaging procedures including radiation exposure.

The imaging environment should be oriented toward PT and practiced in a way that enhances respect, dignity, integrity, and confidentiality of PT. The imaging facilities and individual's professional capacity and ability to deliver patient-focused medical imaging services require competency, confidence, skills, and knowledge. Medicine has recently undergone a paradigm shift of added value in the delivery of services in achieving the desired results using the most cost-effective use of resources while sustaining integrated approach to a patient-centric approach of care (Papp, 2015). In other words, any intervention or action requires a sustainable system capable in providing an infrastructure such as facilities and equipment, which goes hand-in-hand with the appropriate workforce, and be innovative and responsive to the PT needs (Lau, 2007). The reality is that people come to the health facility to seek validation of their health status because there is a form disruption of their day-to-day activities. They are unable to detect the source of the problem and are seeking answers to be treated. Structural proced-ures should be in place to monitor the quality of investigations, human resources, hospital activities including management, education, and training of staff, and the spatial and temporal logistics of PT if affiliated to a healthcare complex.

THE DISTRIBUTED NATURE OF IMAGING SERVICES

The referral for imaging investigations and the imaging itself is 'in the middle of the continuum' (Boon, et al., 2009, p. 716). A discrete facility depending on the level of care service may include fully integrated medical imaging services or most times co-located interventional imaging services, nuclear medicine, and radiation therapy units. The model of service delivery could provide as a single department managed and operated by the health institution, or a main facility with subunits located for ease of PT access within the hospital. Privately owned and/or shared public-operated imaging facility provides a service to several hos-pital units on-site and off-site to referring providers on an outpatient's approach.

Within this context, QM can be defined as a set of activities that occur within and between the medical services available across the different levels of the healthcare system (Charles, et al., 2006; Van Rensburg and Pelser, 2004). As previously acknowledged, these activities require interactions with a diverse set of healthcare practitioners. This being the case, the imaging procedure requires both the time and the space to occur. The permanent stability of an organizational and physical structure facilitates the timely delivery of the imaging procedure for both referrers and PTs in terms of equitable access.

INITIATION OF THE REQUEST WITHIN THE MEDICAL ENCOUNTER: 'X-RAYS, SO YOU ACTUALLY LOOK INTO THEIR PROBLEMS?'

There is a dearth of literature on overutilization of imaging services and safety in relation to health outcomes. The principles of justification and optimization that underpin medical imaging practice also form the cornerstones of radi-ation protection (Malone, 2012). Referring practitioners are expected to employ the most efficient diagnostic strategies to prevent unnecessary referrals for diagnostic tests that could impact on time and resources (Kravitz and Call-ahan, 2000). Issues such as the 'placebo effect' (Stoddart and Holl, 1989) of an image investigation are related to the provider's perception of a PT who regards a referral for a diagnostic imaging investigation as a curative measure to the extent of healing their illness. For the so-called 'demanding PT' to appease their desire for an imaging investigation, appears as if the doctor is looking into their problem and an incentive to prevent patient dissatisfaction (Broder, et al., 2016; Watson, et al., 2017). 'Or it's a belief that it's happening (a tangible outcomes)' [S1] and the PT 'is content, the[re] is progress of their treatment' [S1] (Krestin, 2010). At times imaging investigations reassure that something is being done to detect what is wrong with them. Referrals could be initiated 'out of moral obligation' [S1]. This creates a false sense hope for the PT whereby that something could and have been done. A nonreferral of an imaging investigation is maybe perceived as the possibility of missing 'what is wrong with them' [S1].

From a psychosocial perspective, to avoid friction despite the risk and cost is because their peers told them:

> ... you fight with them or just comply and write out a request form and send them. Sometimes the easiest way is to comply, even if you know you're not going to show anything. It set their minds at ease and not avoid trouble for yourself and the misconception, that the PT might go away thinking, that lousy doctor did not send me for X-rays ... It is ... psychological that you do something even if it's not 100% indicated because they are worried.
>
> [S1]

Murphy (2001, p. 200) states that myths that confuse a PT 'blur the boundaries between facts and fiction' and are widely disseminated as a result of a PT previous encounters with imaging examinations.

According to Geneau, et al. (2008), a PT can influence a physicians' behavior. Espeland and Baerheim (2003) found that strong wishes from PTs in cases where the clinical indication for radiographic investigation is in doubt, little else could be done, the consultation is difficult, or time is scarce:

> In the outpatient settings one has the time to explain that. But in an emergency setting you don't have the time, where every minute counts, when you're seeing hundreds of PT a day. That 5 or 10 minutes in an emergency setting is very valuable.
>
> [S1]

It is argued that patients arrive for diagnostic investigations with different levels of preparedness of what an imaging investigation entails. This could be because it's a *'rush in rush ... couple minutes and out'* [S2]. For instance, *'The doctor just looks at my foot [leg] and he say, 'I must go to the X-ray and bring it back again'* [S1]. In some instances:

> PT: *'Well, he is the doctor and usually when you get sort of the symptoms, I've got to, explain to him what's wrong, he knew straight away. Righto, [he] look behind me.*
> Doctor: *'I want to check your neck'*
> PT: *Started pressing me [neck],*
> Doctor: *'that hurt, that hurt, that hurt'*
> PT: *Nah, nah, [then when he hit it],*
> Doctor: C7
> PT: *He didn't hesitate, right before you know, he sent me for an x-ray' ... So, the main aim with help of the doctor to be treated and managed ... for me to move on*
>
> [S2]

With no information, the assumption, *'In general radiography, ... the doctor does not explain the chest X-rays and all the details to the PT presuming that the PT knows about it'* [S1]. From a radiology nurses' perspective, it is important, *'To reassure the PT. Do you know why you are here? Do you know what is expected? Once more just to make them feel better'* [S1].

Additionally:

> ... To have well-informed PT ... situation rather than having PT not knowing what is going on. Unfortunately, ... there is ... big communication barrier ... with PTs that makes it difficult to communicate exactly what is wrong without there been confusion in the PT mind.
>
> [S1]

MEDICAL ENCOUNTER: 'BECAUSE WE DON'T TALK TO THE PT ABOUT IT' [S1]

According Malone, et al. (2012) and Portelli, et al. (2018), communication of what the investigation entails, frequently little or no mention is made of radiation risk.

> – That is where the doctors are lacking a lot. ... the risk and it is dangerous ....
>
> [S1]

The assumption is that PTs are knowledgeable and have insight on general radiographic investigations and the associated risks. That is detailed a little further:

> I never sort of discuss the benefits as such. It's for me logical that the benefit is that we will see that there is a fracture. I understand it might be beneficial to tell them. I just feel that in a setting like that they already know why I am sending them.
>
> [S1]

These situations are complicated by the complex and specialized nature of the units used to quantify radiation exposure, which is not conducive to effective communication with the public and even with health professionals:

> I think it is a good theory PT having a small card with a record of their radiation doses] but I think the difficulty will come in implementing and get[ting] the PT up to a level of understanding where he/she understands that I have been exposed to this many rads of radiation whichever measurements you want to use and I can take only a certain amount more. I think you have to educate the PT to such a level that they understand risk versus benefit; I don't how you're going to implement it.
>
> [S1]

Malone, et al. (2012) state that PTs have the right to know and understand radiation risk and it is the duty of the referrer to inform them. According to radiologists, this contributes to empowering PTs to make informed decisions:

When they discuss with PT, 'This is the situation; these are the options that you have and these are the disadvantages of all the options that you have, including radiation'. And they should, you know, weigh their advantages over disadvantages. And when they come here, they are knowledgeable and quite understanding.

[S1]

The reality, – '... *we even forget to ask them if they are pregnant*' [S1]. From the PT perspective, they may be aware, '*because they did not want you to infect the baby. I notice the sign if you are pregnant you must respond to them quickly*' [S1].

Most often PTs adopt a submissive role because of inadequate communication with and orientation of PT who are not familiar with the referral processes and procedures, including what to expect of a diagnostic investigation. PTs opt to ask questions and seek answers from nurses especially in the wards:

Why are we going to X-ray? 'What are X-rays? What are we going to do there? Are there any big machines? Am I going to feel the pain when they're going to do this?'... But in the ward, because they don't know, they have got the fear of maybe I am lying to them, like they are going to feel the pain. And when they come back, I ask, 'How was it?' They say, 'Eish, I am better now because I thought you were lying. It wasn't that difficult'.

[NURSE, DH in Makanjee, et al., 2014, p. 109]

PT questions often tend to draw correlations with procedures or interventions with which they are more familiar, and could lead to expecting the worst., '... *What are they going to do with me? What is wrong? Are they going to operate?*' [S1]. This sense of expectation of a physical experience or a change regarding their physical status could be because the notion of X-rays is an abstract concept and difficult to understand. To the contrary, some PTs adopt a neutral stance and opt '*to go with flow*' [S1].

THE DIAGNOSTIC IMAGING REQUEST: 'YOU KNOW [A] LITTLE BIT ABOUT WHAT'S GOING ON'

Based on the medical encounter, radiographers observe a '*peep*' [S1] of the medical encounter. This poses a dilemma for radiographers on how to fill in the communication gaps without risking the implied consent and relational aspects between them. Further, interactions and decision-making regarding the imaging context is inherently triadic in nature. That is, the radiographers, the PT, and the imaging request by the referring clinician, which is mediated through text (request order digital or paper based) and technology (RIS/HIS and imaging equipment). From a postmodern perspective, both radiographers and radiologist roles and responsibility are sort of seen as the social actors 'bridging the divide between technology and human healthcare' (Murphy, 2006).

THE DIAGNOSTIC IMAGING REQUEST: THE COMPLEX JUDGMENT CALL

The request form plays a central role in all aspects of the systemic interaction process. From the viewpoint of a radiographer's scope of practice, the request form serves as a form of authorization to perform an investigation and as the primary medium of communication between themselves and the referring clinician (cf, Dhingsa, et al., 2002). The quality of information and standards of the completeness on the request order is essential (Rawle and Pighills, 2018). However, request orders may not be sent at the time of booking, can be incomplete, and/or there can be discrepancy or noncorrelation between the information on the request form and the information provided by PT, requiring the radiographer to contact the referrer. This will also depend on their availability especially during handover. In some instances, the radiographer may consult with a radiologist on site or head of the referring unit in the absence of the referring practitioner. Furthermore, radiographers need to take a discretionary judgment call as to whether to reject a request. This leads to PT landing in between space while been shunted 'to and froing' with risk of a delayed procedures.

A decision to proceed with an investigation requested (depends on the completeness of the request), that is, to accept as or amend the required examination/ projection based on the information gathered from PT. Findings from this research suggests that PT may be seldom communicated in this regard. For example, if the radiographer goes ahead, it is essential not to improvise on the quality and safety of the PT. The example below highlights a radiographer reflecting on how an orthopedic registrar omitted the clinical history of a patient and the risk avoided by a 'judgment call' by the radiographer:

> We had a patient ... for a pre-op chest X-ray ... the clinical history ... a pre-op. ... It's something that we do every day ... We moved the patient in one of the bigger rooms so that we can do the patient with a mobile because he said that he couldn't stand. I am standing there and looking at the patient. I don't see any casts, I don't see any traction, I don't see any anything that indicates that there is an extremity fracture. And then the students are already standing there holding the patients arms. They want to lift up so that we can put the cassette in. And then I said, 'Guys, just hold on a second'. And I check the X-ray form and contacted the referring *clinician*:
>
> [S1]

> *Radiographer: 'I just want to find out from you what is wrong with him? Before we lift him up to do the X-ray'.*
> *Referrer: 'Oh no, he has got an unstable fracture of T4'.*
> *Radiographer: '... we could have caused more harm to the patient because you did not fill in the X-ray form correctly'.*
> *Referrer: 'Oh sorry, I forgot'.*
>
> [S1]

Later, the radiographer reflects on this experience:

We have a lot of stuff like that to deal with every day. Because it is an academic hospital, and the professors and the registrars don't always give them adequate education on how to fill in an X-ray form correctly and what must be on it, and specific things that we need to know ... We must try and guess what's wrong and try and not to injure patients further by doing something that we're not supposed to do because of his/her injuries. So we have a really big problem whereby X-ray forms not being filled out correctly.

[S1]

THE DIAGNOSTIC IMAGING REQUEST: 'IT'S DIFFICULT FOR US SOMETIMES TO TRY AND FIGURE OUT WHAT WE'RE ALLOWED TO SAY'

Kelly, et al. (2012) suggest that the accuracy of diagnostic decisions can be improved through collaboration between radiographers and doctors. However, power of relations in the hierarchies leaves very little space for radiographers to participate in decision-making processes, which may be dominated by the medical professionals. It could be argued that medical professionals see the completion of the request form as a medium of 'instruction' for radiographers in order to perform a job:

My requesting may be unreasonable, but still, if I, on my clinical decision-making, want a specific view or specific X-ray, I don't think they can refuse. So that is my feeling; that is my biggest thing ... Where people are not informed properly, again that's the relation between the doctor and their patient. That's almost like a triangle between the doctor, the patient, and then the radiographer. ... The radiographers are somehow are involved, that they're involved in the process to the benefit of the patient or sometimes not to the benefit [of the PT]. You don't make the wrong comment about something. They jeopardise the relationship between the doctors. If they, let's say for example, I see the patient, I send him for X-ray. The radiographer takes my request form and while she is busy with the patient, she will make the comment like, 'why did the doctor want this X-ray?' Then the PT comes back to you, 'Doctor they say you are sending me unnecessary for the X-ray'. Maybe there is now a little bit friction. So I don't think the radiographer is in a position to always comment or make a decision about the requests made by doctors.

[S1]

In other instances, *when you even call the doctor sometimes, they are very rude. They just tell you, 'just do the X-rays'.* [S1]

In these types of circumstances, a skilled evidenced-based assessment with a sound judgment and decision is called for. The needs expressed above resonates with Croskerry's (2005) notion of situational awareness – knowing what has gone before, what is happening now, and anticipating what is coming now. Whatever the decision, the radiographers need to provide comprehensive written accounts on the incident.

Protocols and policies in this regard are helpful, such as an audit trail and incorporation in part of continuous professional development part of the QI process; having ongoing discussions between the various healthcare professionals regarding their expectations regarding referrals. Device an informed medical imaging referral guidelines, protocols, and policies could be a solution. An integral part of this process is to bear in mind the reality that the medical imaging service is distributed in nature and requires different types of services at one given point in time. Lack of resources could lead to mistrust between professionals, thereby putting the safety of patients at risk.

ACTUAL IMAGING INVESTIGATION ENCOUNTER: 'THE COMMUNICATION PART ITSELF MAKES IT COMFORTABLE, NOT THE ACTUAL THING'

From the research findings, it can often vary from radiographer to radiographer. Interactions between radiographers and PT are often offer (a) no interaction, (b) download of information on the PT, '*the student was explaining everything she was doing*' and/or (c) equal exchanges between the radiographer and the patient. In a recent study, the interactions were oriented to ease tensions when students undertook investigation as part of their skills development:

> … I mean the young girl that was in there just now who was there for work experience was full of life, bright. I walked in there and was taken away like hello, who's this and she says I'm work experience and tells me that she's down at Sutherland hospital and the other lady [Supervising Radiographer] in there I said, 'Is she cracking the whip?'. So I made a joke you know ….
>
> [S2]

Engagement regarding the investigations of patients found it easier to 'talk to and gel around'. The radiographer shared: '*He had pains and he showed me*' [S1]. This information was verified on the request form – '*I read the history again and it made sense*' [S1]. Figure 8.1 depicts the types of interactions that may occur between radiographers and PT.

ACTUAL IMAGING INVESTIGATION ENCOUNTER: 'NICE AND EASY, NO DRAMAS AT ALL'

Engaging with PTs goes beyond verbal communication and arguably involves a nurturing component (Freeman-Sanderson, et al., 2018), '*… unwell, injured or vulnerable you do want to know that people are looking after you. Showing a bit of interest and care into you*'.

[S2]

Radiographers are generally, described by patients as '*girls*' or '*boys*' that are '*well mannered*'. Unlike the accustomed impersonal radiographers:

> They didn't introduce themselves, … assumed I knew her name … we met the last time … the student introduced herself.
>
> [S2]

Seldom creating a space for the PT voice, to seek clarity, or asking questions before the procedure, '*… if everything is okay*' [S2] and actively engage in

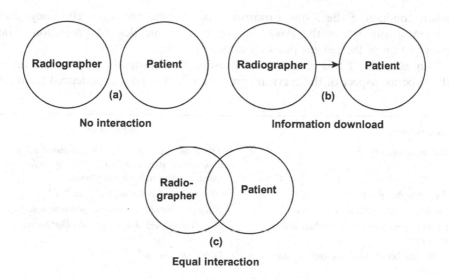

FIGURE 8.1   Patterns of interaction during information gathering and verification.

a two-way interactions regarding the unfamiliar environment (O'Connor and Halkett, 2018). It is argued that the PT must also be knowledgeable on the investigations and be able to ask questions (Rajala, et al., 2018). Throughout these interactions, it is important to consider what the patient presents with, namely in an emergency context. Like the medical practitioners, radiographers were also accustomed to the technological procedures as PT would be familiar with imaging processes – *'And sometimes I know it's because you're doing it often, but they come across as being bored'*.

[S2]

Essentially the focus shifts to a stochato cues of information and instructions with a view of getting physical cooperation in terms of orientating the body part for a specific projection or projections as in Figure 8.1(b). While the patients adopt an approach to give the healthcare providers not only when undergoing an investigation but throughout the processes involved acquiring an imaging investigation a chance to do their job *'I kind of just calm myself and get it over with, get it done …'*.

[S2]

THE ACTUAL IMAGING ENCOUNTER: 'WHAT'S GOING ON?' VERSUS 'NOT MUCH YOU CAN DO ABOUT IT'

For radiographers and radiologists, to be task orientated is important in acquiring optimal image quality, but it is equally important to keep PT informed of what is happening and take the PT construction of the understanding of the event into account (Murphy, 2001). It can be useful, for example, to demonstrate what is required before positioning PT. *We have to do like step by step'* [S1]. Of importance is the sharp skill of remaining vigilant when conducting the X-ray exposure. That is, the precision, efficiency, and accuracy with which the image is captured of the desired projection. The cooperation of the patient plays an important role.

An example of this includes an open-mouth investigation, which examines the superior aspect of the cervical spine for PTs that have experienced trauma:

Observation	Interaction
Radiographer explains.	*'Okay, this one is a bit strange, because I ask you to open your mouth so that we can see the first vertebra. … But I will tell you when'.*
Radiographer positions.	*'Just want to feel the back of your head'.*
Radiographer goes ahead and feels.	*'Keep your head like this and open your mouth'.*
Patient opens mouth. The radiographer appraises.	*'That's it, perfect. Just hold it like that for me'.*
Radiographer stands back and irradiates the patient.	*'Say aaahh for me'.*
While patient is making the sound, he moves the head and stops while the radiographer exposes.	*'Okay, you can relax'.*
Patient closes mouth and radiographer views the image, is not satisfied, and decides to do another one.	*'I just want to get a better one'.*
	[S1]

Equally important is to critically think and problem solve with confidence to conduct the investigation with minimal discomfort to a patient. A typical example in the absence of field-expert support, the referrer may decide to 'clear' the cervical spine following trauma by manipulating the patient, that is, pulling arms inferiorly to visualize the cervical and thorical spine junction. While this remains a collaborative effort in the clinical setting in obtaining sound radiographic images, there are examples where insufficient knowledge can lead to challenges between radiographers and referrers:

With some, it's just impossible to get it [the desired projection]. For example, if a patient comes for a whole spine X-ray on the stretcher I know I can do that, but then they ask for a lateral shoulder. You have to turn the patient, which is unsafe. Then you end up not doing those X-rays. You send the patient back and then the doctor will ask you, she asked for a lateral shoulder, why didn't you do that? Then you must explain to the doctor that you first must clear the cervical spine before we can move the patient to obtain a lateral projection. 'Oh, I didn't think about that'. Then you end up, how does this happen? How does the doctor not think about it?

[S1]

Throughout these interactions, it is important to consider what the patient presents with, namely in an unpredictable emergency context. As illustrated in the following quote, which may result in a temporary bottle necking but justi-fied due to systemic matters where the care becomes lean: *'Yeah, I think also the day that it happened the room was full and it everyone was in a hurry and they needed the machine to do another PT'* [S2].

POST INVESTIGATION: 'THE QUICK AND EASY'

After the investigation, the radiographer affirms, *'Sit there outside'*. It is gener-ally accepted that PTs adopt a pattern of compliance, due to an uncertainty of what is expected of the outcome. *'Why was I there, nothing happened to me?'* [S1]. Therefore, PTs often share a sense of relief after the investigation, fre-quently using an expression like *'I really got in quickly ...'* [S2]. Providers often construct PT expression of relief from anxiety as a misconception of feeling better as part of a healing process.

OUTCOMES OF THE INVESTIGATION: 'US' AND 'THEM' – THE SPACE FOR THE PT VOICE

PTs do not normally question the referrer on the outcomes of an investigation because they were, according to nurses: *'scared to ask the doctor'* [S1]. Edwards and Sines (2007) identified in their study that PTs were often 'passive within the decision-making process, merely acting as an unreliable, and some-times reluctant, repository of information' (p. 2451).

For general radiographic imaging, there are two types interactional pat-terns. Figure 8.2(a) reflects the relationship mediated by technology, some-times with the interaction of a radiographer playing a mediating role with PT. Figure 8.2(b) depicts a situation that includes the PT throughout the process.

Interactions between radiologists, radiographers, and PTs are characterized by different levels of engagement that are, to a certain extent, dependent on individuals and their styles of communication. If a radiologist is on site, radio-graphers are likely to clarify uncertainty with radiologists regarding an investi-gation. For example: *'... We're not sure about this. Can you just look at the shoulder? Do you think we need another view?'* [S1]. The type of information shared among healthcare professionals forms an integral component in build-ing relations and trust with PT and attempts to have minimal disruption:

> ... Sometimes we tell the radiographer or discuss with the radiographer. They are the ones who talk to the patients afterwards, especially when it comes to after the procedure. So, they're quite important in providing a go-between or mediator's role.
>
> [S1]

Patients expressed the desire of integrating the interpretation of the image and the role of the diagnostic procedure with radiographers in order for them to identify what treatment and/or management they may need:

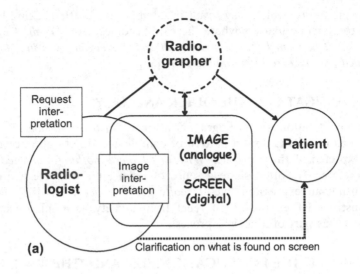

(a)

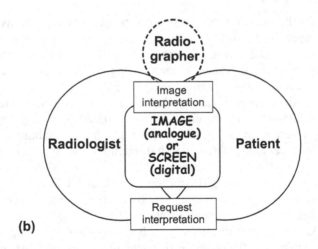

(b)

FIGURE 8.2 Patterns of interactions regarding outcomes of a diagnostic imaging investigation.

Because I don't know how they read them, I would really not know. … I would have liked them to show me, to say, 'If you see this and that picture, it means you have to take this' [medication]. I can take, even if I don't understand.

[S1]

Some PTs demonstrate active involvement by showing interest:

Yah, the thing which I am crying about, is it getting worse? They compared those two X-rays, it must be some three months back and the other one is now that they will compare. The way I see those X-rays, I am satisfied they are still showing the same.

[S1]

This gives PT a sense of control and understanding over what is happening (Borgen, et al., 2010). In most instances, the patient feels like a mere bystander when discussions were being held about them and not with them. To raise questions of concerns becomes bleak.

The doctors sometimes they will come with the x-rays in front of the PT. They talk among themselves as a team. They don't explain to the PT. This is your x-rays, 'This is what we found, see this and this, this is happening to you'. So, if the doctor can talk to the PTs and come to the level of the PT that will help, because PTs are afraid of asking the doctors.

[S1]

In some instances – 'they didn't explain anything before or after. Following the X-rays they didn't tell me anything. I just had to find out for later' [S1].

There are also instances where the interpretation of a follow-up radiographic image varied (Mattsson, et al., 2018):

They told me two days ago my child's bone is not growing. The X-rays show it's not growing at all, this doctor who told me. And now yesterday they [another doctor] told me, no, but the bone is growing. Now how do you explain that? The one-day they tell you there is no difference. She's been lying here for one month. And the next day the doctor says the bone is all right; it is starting to grow. ... Then he told me they looked at the X-rays well ... the bone is busy growing, bone formation, but not quite as it should be 100%. So how did they see it if the first doctor had said the bone is not growing? I don't know what has happened there. Because the one doctor says this and the other says that. So you are so confused ...

[P1 in Makanjee, et al., 2015, p. 202]

OVERARCHING CHALLENGES OF A HOLISTIC APPROACH TO CARE: 'WHAT AM I GOING TO DO AND WHERE AM I GOING TO GO IF I DIDN'T COME'

Earlier in the introduction, the concept of value-added care was highlighted whereby patient care within the imaging setting is wider than just the investigation itself. So, patient centeredness and person-centered concepts are interrelated to the psychosocial domain. PTs express a variety of feelings and emotions, such as the feeling of uncertainty, negativity, and anxiety, but also a sense of hope and then relief and even optimism after learning the outcomes of their investigation.

I am just waiting for good results. I don't want to lose my leg. I am a domestic worker, a single parent with two children ... I have not arranged with anyone to look after them and am already two days away from work ...

[S1]

Confounding issues in a fragmented environment is illustrated in the following quote with a mismatch in expectations and effective communication of outcomes.

> ... So maybe the doctor didn't, or we didn't understand each other very well. She asked me questions and I answer and it's like she didn't see the seriousness that I am going through. I can't say I wasn't happy, but I wanted to see after the X-rays how it goes. Today is Monday, but unfortunately she wasn't here today.
>
> [S1]

ISSUES OF DISPARITIES: 'THE PATIENT DOES NOT UNDERSTAND EVERYTHING THAT YOU ARE SAYING'

When information is passed on, account should be taken of an individual's language preferences and level of literacy (Leung, et al., 2018) including their ability to negotiate and coordinate care (Suter, et al., 2009). PT may feel more comfortable if communicated to in their language of preference. Seldom do PT alert the attending health professional early enough of their language preference. Because of a judgmental approach often results in the integrity and dignity of PT are undermined.

> Sometimes they [the doctors] will call you and you must come and interpret for them. And then they don't know how far that patient, she has got knowledge. They will call you, even if you are nowhere, 'Hey you, will you please come and interpret!' And then I told them, 'Just first ask the patient, can you talk English or Afrikaans?'.
>
> [Nurse, DH in Makanjee, et al., 2014, p. 109]

One way of overcoming this is to include language preferences on request forms. Confirm with PT upon arrival their ability to communicate in a specific language. Often family members are involved in the care process; they also are the preferred interpreters. One is cautioned against information getting '*lost in translation*' and the patients may not understand. The use of a professional interpreter could be received with some resistance because of issues linked to privacy, cost, confidence, and trust (Laidsaar-Powell, et al., 2018). It would be appropriate to address the reasons and emphasize the value of an accurate interpretation in supporting the patient to avoid misunderstandings. To bridge this gap is the ability to use comprehensive language. Modify some of the terminology through using metaphors (Dao, et al., 2018; Van Ravesteijn, et al., 2012).

> Some don't even know what X-ray [is]. They ask you, 'Am I going to feel the pain when they're going to do this?' And then I try to explain to them because I saw some chest X-rays being done before. I will tell her that you are going to stand there and they have got a big light and a big machine and they are going to take a photo of you. But this photo [is] take[n] of your inside, not your outside appearance.
>
> (Nurse, DH in Makanjee, et al., 2014, p. 109)

For the PT to understand what is going on requires an interaction pitched at a level where the PT could engage effectively (cf. Henderson, 2006). One of the challenges is the issue of conveying too much information for PT to digest. PTs are expected to keep up with the pace of communication while constructing their understanding and meaning making and be in control of their care and treatment (Joseph, et al., 2018).

CONCLUSION

This chapter directly and indirectly involves accounts of patients and the patient's perspectives regarding their expectations and experiences. Further, it highlights that to sustain a valued quality diagnostic imaging service and offer holistic care, can be complex. Interactions focusing on task-oriented communication styles with little room for two-way interactions may still predominates in the healthcare setting. From a patient perspective, as soon as the imaging investigation is over the 'looming uncertainty' as they wait for an outcome. The value and the worth of the investigations are reflected in the triadic encounter, namely the image/s and the accompanying radiological report. This, together with the practitioner, enables discussion of treatment and management plan for the patient. A paradigm shift is required regarding the 'isolated QI management' of diagnostic imaging services to an integrated healthcare approach as part of the continuum of care process. Including aspects of respect, integrity, and dignity of the patient as a person, which extends beyond the boundaries of the health system, such as the persons quality of life dimensions, and the reliance on family and community support systems in dealing with the anticipated outcomes of a general radiographic image. To achieve a collective QI management improvement requires all of the stakeholders to be involved.

REFERENCES

Abdous, M., 2009. E-learning quality assurance: a process-oriented lifecycle model. *Quality Assurance in Education*, 17(3), 281–295.

Berwick, D.M., 1998. Developing and testing changes in delivery of care. *Annals Internal Medicine*, 128, 651–656.

Boon, H.S., Mior, S.A., Bamsley, J., Ashbury, F.D. and Haing, R., 2009. The difference between integration and collaboration in patient care: results from key informant interviews working in multiprofessional healthcare teams. *Journal of Manipulative and Physiological Therapeutics*, 32(9), 715–722.

Borgen, L., Stranden, E., Espeland, A. (2010). Clinicians' justification of imaging: do radiation issues play a role. *Insights Imaging, 1(3)*, 193–200.

Broder, J.S., Bhat, R., Boyd, J.P., Ogloblin, I.A., Limkakeng, A., Hocker, M.B., Drake, W.G., Miller, T., Harringa, J.B. and Repplinger, M.D., 2016. Who explicitly requests the ordering of computed tomography for emergency department patients? A multicenter prospective study. *Emergency Radiology*, 23(3), 221–227.

Charles, C., Gafni, A., Whelan, T. and O' Brien, M.A., 2006. Cultural influences on the physician PT encounter: the case of shared treatment decision-making. *Patient Education and Counseling*, 63(3), 262e7.

Croskerry, K., 2005. The theory and practice of clinical decision-making. *The Canadian Journal of Anaesthesia*, 52(1), R1–R8.

Dao, M.D., Inglin, S., Vilpert, S. and Hudelson, P., 2018. The relevance of clinical ethnography: reflections on 10 years of a cultural consultation service. *BMC Health Services Research*, 18(1), 19.

Dhingsa, R., Finlay, D.B.L., Robinson, G.D. and Liddicoat, A.J., 2002. Assessment of agreement between general practitioners and radiologists as to whether a radiation exposure is justified. *British Journal of Radiology*, 75(890), 136–139.

Edwards, B. and Sines, D., 2007. Passing the audition – the appraisal of client credibility and assessment by nurses at triage. *Journal of Clinical Nursing*, 17(18), 2444–2451.

Espeland, A. and Baerheim, A., 2003. Factors affecting general practitioners' decisions about plain radiography for back pain: implications for classification of guideline barriers – a qualitative study. *BMC Health Services Research*, 3(8), 1–10.

Freeman-Sanderson, A.L., Togher, L., Elkins, M. and Kenny, B., 2018. Quality of life improves for tracheostomy patients with return of voice: a mixed methods evaluation of the patient experience across the care continuum. *Intensive and Critical Care Nursing*, 46, 10–16.

Geneau, R., Lehoux, P., Pineault, R. and Lamarche, P., 2008. Understanding the work of general practitioners: a social science perspective on the context of medical decision making in primary care. *BMC Family Practice*, 9(12), 1–10.

Harvey, H.B., Hassanzadeh, E., Aron, S., Rosenthal, D.I., Thrall, J.H. and Abujudeh, H.H., 2016. Key performance indicators in radiology: you can't manage what you can't measure current problems in diagnostic. *Radiology*, 45, 115–121. doi: https://doi.org/10.1067/j.cpradiol.2015.07.014

Henderson, A., 2006. Boundaries around the 'well-informed' patient: the contribution of Schutz to inform nurses' interactions. *Journal of Clinical Nursing*, 15(1), 4–10.

Hogden, A., Greenfield, D., Brandon, M., Debono, D., Mumford, V., Westbrook, J. and Braithwaite, J., 2018. How does an accreditation programme in residential aged care inform the way residents manage their healthcare and lifestyle?. In Managing Improvement in Healthcare (pp. 295–310). Palgrave Macmillan, Cham.

Joseph, G., Lee, R., Pasick, R.J., Guerra, C., Schillinger, D. and Rubin, S., 2018. Effective communication in the era of precision medicine: a pilot intervention with low health literacy patients to improve genetic counselling communication. *European Journal of Medical Genetics*, S1769–7212(18), 30457–30459. doi: 10.1016/j.ejmg.2018

Kelly, B.S., Rainford, L.A., Gray, J. and McEntee, M.F., 2012. Collaboration between radiological technologists (radiographers) and junior doctors during image interpretation improves the accuracy of diagnostic decisions. *Radiography*, 18(2), 90–95.

Kravitz, R.L. and Callahan, E.J., 2000. Patient's perceptions of omitted examinations and tests: a qualitative analysis. *Journal of General Internal Medicine*, 15(1), 38–45.

Krestin, G.P., 2010. Commoditization in radiology: threat or opportunity? *Radiology*, 256(2), 338–342.

Lagrosen, S., Seyyed-Hashemi, R. and Leitner, M., 2004. Examination of the dimensions of quality in higher education. *Quality Assurance in Education*, 12(2), 61–69.

Laidsaar-Powell, R., Butow, P., Boyle, F. and Juraskova, I., 2018. Managing challenging interactions with family caregivers in the cancer setting: guidelines for clinicians (TRIO Guidelines-2). *Patient Eucation and Counseling*, 101(6), 983–994.

Lau, L.S., 2007. Leadership and management in quality radiology. *Biomedical Imaging and Intervention Journal*, 3(3), e21.

Leung, L.B., Vargas-Bustamante, A., Martinez, A.E., Chen, X. and Rodriguez, H.P., 2018. Disparities in diabetes care quality by english language preference in community health centers. *Health services research*, 53(1), 509–531.

Lovric, G.T. and Makanjee, C.R., 2018. A qualitative study exploring patients' expectations and experiences of the localization event as part of radiation therapy. *Journal of Radiology Nursing*, 37(3), 205–210.

Makanjee, C.R., Bergh, A.M. and Hoffmann, W.A., 2014. "So you are running between" – a qualitative study of nurses' involvement with diagnostic imaging in South Africa. *Journal of Radiology Nursing*, 33(3), 105–115.

Makanjee, C.R., Bergh, A.M. and Hoffmann, W.A., 2015. Multiprovider and patient perspectives on conveying diagnostic imaging investigation results in a South African public health care system. *Journal of Medical Imaging and Radiation Sciences*, 46(2), 197–204.

Malone, J., Guleria, R., Craven, C., Horton, P., Järvinen, H., Mayo, J., O'Reilly, G., Picano, E., Remedios, D., Leheron, J., Rehani, M., Holmberg, O. and Czarwinski, R., 2012. Justification of diagnostic medical exposures, some practical issues: report of an International Atomic Energy Agency Consultation. *The British Journal of Radiology*, 85(1013), 523–538.

Mattsson, B., Ertman, D., Exadaktylos, A.K., Martinolli, L. and Hautz, W.E., 2018. Now you see me: a pragmatic cohort study comparing first and final radiological diagnoses in the emergency department. *BMJ Open*, 8(1), e020230. doi: 10.1136/bmjopen-2017-020230

McGovern, M., Dent, K. and Kessler, R., 2018. A unified model of behavioral health integration in primary care. *Academic Psychiatry*, 42(2), 265–268.

Murphy, F., 2001. Understanding the humanistic interaction with medical imaging technology. *Radiography*, 7(3), 193–201.

Murphy, F.J., 2006. The paradox of imaging technology: a review of the literature. *Radiography*, 12(2), 169–174.

Ngoya, P.S., Muhogora, W.E. and Pitcher, R.D., 2016. Defining the diagnostic divide: an analysis of registered radiological equipment resources in a low-income African country. *The Pan African Medical Journal*, 25(99), 1–7.

O'Connor, M. and Halkett, G., 2018. A systematic review of interventions to reduce psychological distress in pediatric patients receiving radiation therapy. *Patient Education and Counseling*, 102(2), 275–283.

Pahade, J.K., Trout, A.T., Zhang, B., Bhambhvani, P., Muse, V.V., Delaney, L.R., Zucker, E.J., Pandharipande, P.V., Brink, J.A. and Goske, M.J., 2018. What patients want to know about imaging examinations: a multiinstitutional us survey in adult and pediatric teaching hospitals on patient preferences for receiving information before radiologic examinations. *Radiology*, 287(2), 554–562.

Papp, J., 2015. *Quality management in the imaging sciences*. 5th ed. St. Louis, MO: Elsevier.

Portelli, J.L., McNulty, J.P., Bezzina, P. and Rainford, L., 2018. Benefit-risk communication in paediatric imaging: what do referring physicians, radiographers and radiologists think, say and do? *Radiography*, 24(1), 33–40.

Rajala, M., Kaakinen, P., Fordell, M. and Kääriäinen, M., 2018. The quality of patient education in day surgery by adult patients. *Journal of PeriAnesthesia Nursing*, 33(2), 177–187.

Rawle, M. and Pighills, A., 2018. Prevalence of unjustified emergency department x-ray examination referrals performed in a regional Queensland hospital: a pilot study. *Journal of Medical Radiation Sciences*, 65(3), 184–191.

Schout, G. and de Jong, G., 2018. The weakening of kin ties: Exploring the need for life-world led interventions. *International Journal of Environmental Research and Public Health*, 15(2), 203.

Stoddart, P.G.P. and Holl, S.G., 1989. Radiology is valuable to general practitioners; but who pays? *Clinical Radiology*, 40(2), 183–185.

Suter, E., Arndt, J., Arthur, N., Parboosingh, J., Taylor, E. and Deutschlander, S., 2009. Role understanding and effective communication as core competencies for collaborative practice. *Journal of Interprofessional Care*, 23(1), 41–45.

Taylor, M.J., McNicholas, C., Nicolay, C., Darzi, A., Bell, D. and Reed, J.E., 2014. Systematic review of the application of the plan–do–study–act method to improve quality in healthcare. *BMJ Qual Saf*, 23(4), 290–298.

Undeland, M. and Malterud, K., 2002. Diagnostic work in general practice: more than naming a disease. *Scandinavian Journal of Primary Health Care*, 20(3), 145–150.

Van Ravesteijn, H., Van Dijk, I., Darmon, D., Van Der Laar, F. and Lucassen, P., 2012. The reassuring value of diagnostic tests: a systematic review. *Patient Education and Counseling*, 86(1), 3–8.

Van Rensburg, H.C.J. and Pelser, A.J., 2004. The transformation of the South African health system. In: Van Rensburg, H.C.J., Benator, S.R., Doherty, J.E., Heunis, J.C., McIntyre, D.E., Ngwena, C.G., et al., editors. *Health and health care in South Africa*. 1st ed. Pretoria: Van Schaik; p. 149e50.

Watson, J., de Salis, I., Banks, J. and Salisbury, C., 2017. What do tests do for doctors? A qualitative study of blood testing in UK primary care. *Family Practice*, 34(6), 735–739.

Zygmont, M.E., Itri, J.N., Rosenkrantz, A.B., Duong, P-A.T., Gettle, L.M., Mendiratta-Lala, M., Scali, E.P., Winokur, R.S., Probyn, L., Kung, J.W., Bakow, E. and Kadom, N., 2017. Special report: radiology research in quality and safety: current trends and future needs. *Acadamic Radiology*, 24(3), 263–272.

9 General Radiography in the Critical Care and Trauma Environment

Tom Campbell-Adams

THE ROLE OF CONVENTIONAL IMAGING IN CRITICAL CARE

Critical care refers to any care setting primarily for patients with potentially life-threatening conditions in need of constant monitoring and one-to-one care. Commonly, patients will be sedated and ventilated, with some specialist centers offering specialist life-supporting interventions, such as extracorporeal membranous oxygenation. Due to the unstable and seriously ill nature of the patients in critical care settings, it is rare for a patient to be able to travel to the department for imaging and, therefore, the vast majority of imaging will be conducted at the bedside of the patient using a portable X-ray machine.

The most obvious example of a critical care setting is the intensive care unit (ICU) and associated step-down high-dependency unit (HDU), but this is not the only example of a critical care ward. Other forms of critical care include intensive or high-dependency cardiac units (often referred to as critical cardiac units), 'Recovery' wards set aside for patients following anesthesia and surgery, and any form of HDU for a given branch of medicine. Additionally, there are specialist pediatric critical care settings, such as the pediatric intensive care unit (PICU) and neonatal intensive care unit (NICU).

Conventional imaging, and in particular chest radiography, plays an important role in the diagnosis and monitoring of patients in critical care settings. Chest radiographs are often requested to diagnose and monitor resolution of pulmonary edema or effusion or other critical chest pathology such as pneumothoraces (particularly following thoracic surgery). Often, pathology is a result of acute respiratory distress syndrome or acute cardiac failure (Henschke, Yankelvitz, Wand, Davis, & Shiau, 1996). Imaging will also be requested to monitor the placement of essential lines and catheters including central venous catheters (CVCs), peripherally inserted central catheters (PICC lines), and endotracheal and naso-gastric (NG) tubes.

INTERPROFESSIONAL COMMUNICATION IN CRITICAL CARE

Critical care can be a daunting setting for a newly qualified or otherwise inexperienced radiographer for a number of reasons. First, there are the difficulties in working away from the main department and the support available from colleagues. Secondly, the alien nature of the critical care setting, in which a radiographer must carefully work around unfamiliar life-supporting equipment, and finally, the serious nature of patient conditions in the setting can easily invoke some degree of anxiety and fear of causing harm during positioning of either equipment or the patient.

Fortunately, all the aforementioned concerns are perfectly possible to overcome with sound communication with the nursing staff in the critical care setting. As noted earlier, chest radiography remains a commonly used examination in monitoring patients and therefore radiographers are not necessarily strangers to critical care units. The nurses in charge of patients in the critical setting will be aware of the requirements of imaging and willing to support the radiographer to that end. Additionally, good teamwork and communication between members of different health professions has been proven to lead to an increase in patient safety (Brock et al., 2013).

Prior to any attempt to begin setting up equipment for imaging, it is essential the radiographer introduces himself or herself to the nurse in charge of the patient, explains what support is required, and how the patient needs to be positioned and identify any likely difficulties or barriers to undertaking the examination. As an example, it may be the case that the patient cannot be sat in an erect position. It is important in this situation that the radiographer is receptive to any information given by the nurse, and recognizes when the patients condition may preclude planned patient positioning. In this instance, further discussion will be warranted to ensure that the patient is positioned appropriately to ensure, first, the safety of the patient throughout the imaging process and, secondly, that high quality and appropriately diagnostic images are produced.

The same communication should be used when preparing the exposure, and the radiographer should ensure that all members of staff have retreated to an acceptable distance according to local rules, and/or be wearing appropriate lead equivalent shielding. The radiographer should be aware that often staff and relatives in adjacent bed bays will be in sufficiently close proximity to the X-ray tube to need to retreat further and the radiographer should be appropriately sensitive when requesting that those individuals retreat, bearing in mind in particular that staff may be carrying out sensitive procedures which may not be immediately interrupted.

PATIENT CARE AND IMAGING TECHNIQUE IN CRITICAL CARE

There are several safety aspects of positioning and technique that should be considered in mobile imaging in critical care. Infection control status and precautions should be discussed with nursing staff prior to imaging. Prior to the commencement of any imaging, the radiographer should adopt universal infection control procedure, washing hands with soap and water and wearing appropriate personal protective equipment (PPE) to include, as a minimum, a plastic gown and gloves. Dependent on patient condition, additional PPE may be warranted such as a full-length plastic gown or facemask.

The first issue to consider is the positioning of the mobile X-ray unit itself. Space is often at a premium in critical care units and the machine must be placed appropriately to be able to accurately center for imaging, but in such a way as not to obstruct either staff or equipment. When parking the machine parallel to the patient's bed, it is important to ensure there are no collection bottles for chest drains or other tubing that is at risk of being pulled or damaged due to the position of the machine.

During the positioning of the patient, it is essential the radiographer communicates and works closely with the nursing staff as discussed previously. However, it is important that the radiographer does not assume that they cannot be heard by the patient and should communicate with the patient as they would any other, regardless of level of sedation or the Glasgow Comma Scale (GCS) status. Hearing has been found to be sense that remains in comatose patients, whereby auditory stimulation has been linked to increased levels of consciousness (Tavangar, Shahriary-Kalantary, Salimi, Jarahzadeh, & Sarebanhassanabadi, 2015). Thus, radiographers should ensure the imaging procedure is explained to the patient and reassurance during any movement, particularly where the patient is turned on to their back from a decubitus position, sat upright in the bed, or moved into a supine position. The patient should also be warned of potential discomfort when the imaging receptor is positioned behind the patient.

It is important that any patient movement during positioning is supervised at all times by the nurse in charge of the patient. The radiographer should not assume it is safe to move the patient into either an erect or supine position without first conferring with the nurse. As an example, some patients will be unable to tolerate an erect position for hemodynamic reasons and movement into such a position can lead to a dangerous drop in blood pressure. If during the course of positioning for imaging, the patient has to be moved then particular care should be taken to ensure that any lines or tubing attached to the patient is not disrupted or removed accidentally. Particular care should be given to ensure the endotracheal tube remains in situ during movement and all peripheral or central lines should be accounted for during movement. Often, such lines can become tangled during patient movement which, when unnoticed, can put the patient at significant risk of harm from accidental removal.

When inserting the imaging receptor behind the patient, it is important for the radiographer to protect both the patient from an infection control point of view and themselves, from a manual handling standpoint. In this sense, it is important that the receptor is placed in a protective bag of some form, and many critical care units will have plastic sliding bags specifically for this purpose. By doing so, the radiographer will protect the patient from any additional introduction of infection from the receptor as far as possible and also themselves by making the receptor easier to place and move behind the patient. The radiographer should consider the use of slide sheets to support easy placement of the receptor as well.

Prior to exposure, the radiographer should ensure that, as far as is reasonably practical, any possible artifacts across the area under investigation are either removed or moved in such a way as to minimize artifact on the resultant imaging. As noted before, this should be carried out in discussion with the nurse in charge to identify what leads or other equipment can be removed. It may be the case that not all artifacts can be removed from the area of interest for patient safety reasons, but provided that all efforts are made to remove as much artifact as possible, this is unlikely to be problematic. The removal of artifact is particularly important in any imaging instance where the clinical question is to identify the position of a given line or tube. Failure to remove external artifact may result in an undiagnostic image that will result in additional radiation exposure, which could have been avoided.

In a mobile setting radiation protection, measures are essential not only to reduce dose as far as practicably possible for the patient but also to protect both the radiographer and other individuals within the immediate vicinity of the mobile unit. Fundamentals of good radiographic practice, such as appropriate collimation to the area under investigation, should be employed by the radiographer in order to limit scattered radiation and impede image quality. In mobile chest imaging, a large source-to-image distance should be employed to minimize as far as possible magnification of the heart and to reduce entrance skin dose to the patient. As with chest imaging in department, a high kVp technique should be utilized where possible, although occasionally radiographers will find a lower kVp technique is necessary due to limitations associated with either the mobile unit or type of image receptor. A grid should not be utilized in a mobile setting due to the risk of grid cut off from an inappropriately angled central beam. If using equipment with functionality such as a virtual grid or scatter reducing algorithm, then this should be employed, particularly in patients with above-average habitus.

OPTIMAL PATIENT POSITIONING FOR CHEST IMAGING IN CRITICAL CARE

Consideration should be given to the most appropriate patient position for imaging in the critical care setting. Anteroposterior chest imaging is performed due to the lack of patient compliance and for safety reasons, but a common problem which occurs in anteroposterior chest imaging is the position of the patient and corresponding angulation of the imaging receptor. In all chest imaging, whether the image produced is posteroanterior, anteroposterior, or a lateral, the patient should be in either an erect or supine position. What may commonly be seen in practice within critical care is, due to a patient's inability to be sat in an erect position, a semi-erect image being produced in preference to supine imaging. This is problematic for a number of reasons.

First, the semierect position may be deemed to be undiagnostic where fluid levels are being assessed as this position causes a misleading fluid level meniscus, which means the true amount of fluid within the thoracic cavity cannot be effectively or accurately assessed by the reviewing or reporting clinician. Second, there is no standardization of positioning between examinations carried out by different radiographers or even, potentially, the same radiographer on different occasions. 'Semierect' has no fixed definition and is at the discretion of the radiographer. Therefore, a patient requiring a daily chest radiograph to assess the resolution of an effusion may have a semierect imaging performed at an angle of 60 degrees, 45 degrees, and finally 70 degrees.

Standardization of imaging in critical care is particularly important where images are often requested for the purposes of reviewing the level of resolution of a given symptom, such as an effusion or pneumothorax. Without standardization in patient position, it is impossible for a true comparison to be made between films and in the worst-case scenario, the lack of standardization across films may lead to an inaccurate judgement made on the progression or resolution of disease, leading to less-favorable outcomes for the patient. It is important to note that in anteroposterior chest imaging it almost always preferable to produce a supine radiograph instead of a semierect radiograph.

It is also important during any image, but during imaging for critical care especially, that the radiographer is aware of how the area of interest for a chest radiograph may alter slightly dependent upon the clinical question being asked. For example, for a request to monitor the resolution of a pneumothorax or other chest pathology, a radiograph demonstrating the entire lung fields and soft tissue borders is necessary. However, if the request was, for example, to ascertain the position of an NG tube, the radiographer would be justified in centering the image lower than for a normal chest radiograph in order to ensure that the tip of the tube can be visualized below the left hemi diaphragm. Doing so may mean that some of the apices of the lung fields are missed, but based on the request this would not make the imaging undiagnostic.

NG TUBE EVALUATION

The radiographer has an extremely important role in the evaluation and interpretation of images produced in the critical care setting. This is especially the case in the assessment of the position of NG tube radiographs. NG tubes are inserted to ensure appropriate nutrition for patients who are nil by mouth and are often referred to informally as 'feeding tubes'. A nurse or doctor without visualization of the insertion of the tube inserts NG tubes at the bedside. Imaging is often required to confirm appropriate position of the tube, especially where an acidic aspirate cannot be produced. Failure to confirm appropriate position can result in severe patient harm and death where feeding is commenced while the NG tube is inappropriately placed in the trachea or bronchus.

The NG tube checks should be reported immediately by an appropriately qualified and trained clinician or practitioner as part of hospital policy surrounding the use of NG tubes. As a result of this, some hospitals will have policies in place that NG check imaging should only be performed in normal working hours such as between 08:00 and 20:00 to ensure that images can be reported in an appropriate and timely fashion. There will be occasions, however, when an emergency necessitates imaging outside of normal working hours.

While it is not generally a formal requirement for radiographers to be able to identify the correct position of an NG tube, it is an extremely useful skill to have both in terms of professional development and as an advocate for patient safety. It is fair to say that radiographers should not comment on the correct placement of an NG tube, but provide an opinion where a tube is clearly or likely in an incorrect position. It is also important to note that in any given circumstance, a radiographer opinion does not, and should not, supersede a formal written report from an appropriately qualified clinician or practitioner. However, given the time sensitive nature of imaging for NG tube position and the logistical difficulties that can sometimes arise in reporting of images, particularly during out of hours working, radiographer opinion can be useful from a patient's safety point of view.

Following on from the standard 10-point quality assessment of any image produced, there are four questions to be answered in the evaluation of the position of a NG tube, according to guidance from the National Patient Safety Authority, as follows:

1) Does the NG tube follow the path of the esophagus?
2) Does the NG tube bisect the bifurcation of the bronchi?
3) Does the NG tube remain in the midline until the level of the diaphragm?
4) Is the tip of the NG tube visible below the left hemi diaphragm?

(NHS Improvement, 2016)

If the answer to any of the four questions above is 'no' then the radiographer should ensure that the image is reviewed and reported by an appropriate clinician or practitioner as a matter of urgency. The staff responsible for the patients care should urgently be made aware of the concerns of the radiographer: that the tube is not safe for use and that the report should be reviewed by a ward clinician as soon as possible.

THE ROLE OF GENERAL RADIOGRAPHY IN TRAUMA

Before moving forward, it is important to consider the elephant in the X-ray room – what is the role of conventional imaging for trauma and the acutely unwell patient? It is reasonable to argue that computed tomography (CT) has had an increasing impact for the imaging of patients who have experienced significant trauma. To that end, it is now common to find CT facilities positioned within the emergency department itself, away from the main diagnostic imaging department in many hospitals and particularly in designated major trauma centers.

So, if CT is to be the modality of choice in trauma, and thus the modality of choice in many acute conditions, then what role does conventional imaging retain, if any? To answer this, we can begin by looking to NICE (2016). In their guidelines for the management of major trauma, NICE recommends the use of CT in many instances, especially where multiple injuries or potential cervical spine injury is suspected. Alongside this, however, it is made clear that conventional imaging should be utilized as a part of the primary survey in the Advanced Trauma Life Support (ATLS) process and that conventional imaging should be the first choice for imaging for some injuries, including to the thoracic and lumbar spines.

Ultimately, the speed, convenience, and valuable clinical information that can be gleaned from a radiograph mean that, for now at least, conventional imaging retains a significant role in trauma imaging and imaging of the acutely unwell patient.

THE GOLDEN HOUR

Before considering any aspect of imaging patients who have experience either significant trauma or an acute condition it would be prudent to consider the 'golden hour'. The golden hour is the idea that in cases of significant trauma (or a serious acute condition such as a stroke), a patient is likely to have significantly improved outcome where injuries and other issues are diagnosed and treatment started within approximately 60 minutes. After this time, the likelihood of patient mortality is believed to increase significantly. The golden hour is a somewhat contentious concept within emergency medicine and, in reality, there is limited and occasionally conflicting literature to support the idea itself (Rogers, Rittenhouse, & Gross, 2015).

From a practical point of view, working under the assumption of a 'golden hour' is certainly no bad thing. It is without question that any department providing imaging for an emergency department will be under near constant pressure, with limited rooms and resources for an unpredictable flow of patients. With this in mind, while it can be easy to default to a 'first come, first served' approach to imaging (and this is perhaps more likely where clinicians are pushing to avoid breaking wait time targets), it is critical that clinical need is at the forefront of any prioritization of imaging requests. It is, therefore, imperative that priority is always afforded to patients who have experienced significant trauma or who are otherwise acutely unwell or unstable.

THE ROLE OF THE RADIOGRAPHER IN ATLS

The radiographer plays a crucial role in the emergency department multidisciplinary team. While the use of the traditional trauma series, comprising of a cross table lateral cervical spine radiograph, chest radiograph, and pelvis radiograph, is beginning to be phased out in favor of more comprehensive full body CT scanning, the chest and pelvis radiograph remain an important part of the primary survey in the ATLS setting, particularly where the patient is too unstable to be transferred to the CT scanner.

Within this role, the radiographer has a subtly complex job to perform, not only providing quick and high-quality imaging but also maintaining the controlled area around the X-ray unit and the radiation protection of other members of staff within the area in accordance with the requirements of IRMER (2017). To this end, it is essential for the radiographer to ensure that any episode of ATLS is approached with confidence and authority, but not dictatorially. Successful ATLS imaging is entirely reliant upon effective teamwork, and effective teamwork is reliant upon clear communication.

From a practical perspective, strong communication between the radiographer and clinicians is extremely important in the successful management of ATLS, and begins with basic introductions. Upon arriving to Resus, the radiographer should identify and introduce themselves to the lead clinician, and be prepared to take instruction(s). When directed to do so, the radiographer should position the equipment (either a mobile unit or ceiling-mounted tube) and image receptor appropriately, working around the other members of the ATLS team as appropriate. It is important in this instance to remember that every member of the ATLS team has his or her own job to complete and that this can lead to conflict. In this instance, the radiographer should look to the lead clinician for guidance, and be prepared to either take a step back or politely but firmly request the obstructing member to move to enable quick positioning.

It is central that within any ATLS situation, all radiographic positioning is conducted with minimal patient movement wherever possible. The only time a patient should be repositioned in the trolley is in an instance where the radiographer feels that structures under a trolley may obscure anatomy. These can occasionally be found in the form of steel (or other metallic) struts running under the lateral edges of the radiolucent trolley bed. Assuming both a chest and pelvis radiographs are required, the radiographer should check which is the most clinically urgent and prioritize appropriately. Modern digital radiography (DR) mobile units may allow bedside viewing of images so the order in which images are produced is less important in this instance.

During the exposure phase of radiographic imaging, the radiographer assumes all responsibility for radiation protection. As far as the patient is concerned, fundamentals such as appropriate source to image distance, collimation, and exposure factor selection will ensure that the dose imparted to the patient is as low as reasonably practicable. It is important to remember that in any instance of trauma imaging lead equivalent shielding for the patient should not be used. The radiation protection of other members of the ATLS team can be somewhat more difficult, but it is essential the radiographer ensures all members of the team are either wearing lead equivalent protection or have retreated to an appropriate distance as per local rules (usually approximately two meters). Task fixation may lead members of the team to ignore warnings about exposure and it is therefore essential that radiographers give loud, confident warnings about exposures and direct these warnings to any lingering members of the team. Where these warnings are ignored, often looking to the lead clinician to provide additional authority is sufficient to ensure any remaining staff moves. The lead clinician will ultimately identify which task is most important at any given time and enforce prioritization appropriately.

PATIENT CARE DURING TRAUMA IMAGING IN DEPARTMENT

Not all imaging for trauma or acutely unwell patients will take place away from the department in resuscitation areas or other designated HDUs of the emergency department. Patients will often need additional imaging within the department itself following trauma or an acute episode of illness and radiographers need to be aware of and remain sensitive to additional needs these patients may have during imaging and aspects of patient safety that should be considered.

In any instance following significant trauma or illness, the radiographer should risk assess the request presented and gather further information if appropriate prior to the patient being transferred from the emergency department to the radiographic department. This risk assessment should be to consider whether mobile imaging may be more appropriate due to any number of factors such as hemodynamic stability, low or decreasing GCS, or an infection control issue which may put other patients and individuals at risk. Radiographers should be aware of their local protocols with regards to patient transfer and where mobile imaging can be conducted. Dependent on the patient's condition, it may be necessary to request a chaperone with the patient who may be either a nursing assistant or nurse depending on the severity of the condition.

Prior to commencing any radiographic examination, the radiographer must ensure that they have accessory equipment in place, which may be required during imaging, including a selection of radiolucent pads and supports, which are particularly useful in appendicular skeletal imaging. Depending on the request and patient presentation, it is important to ensure oxygen is available if necessary and that suction is easily and readily available. Suction is especially important in patients who are in a supine condition or under spinal immobilization in order to prevent aspiration of vomit. It is important to be aware that nausea and vomiting can occur as a response to shock from a traumatic incident and as a side effect of analgesia (particularly strong opioids). It is also essential that the radiographer monitors the condition of the patient throughout the imaging procedure as patients will have been stabilized, and in a stable condition prior to transfer, there is often the risk of sudden and unexpected deterioration.

It is important that radiographers communicate fully with the patient throughout the imaging process and maintains a calm and reassuring tone and composure. Patients may be in shock or emotionally upset following trauma or as a result of significant illness; thus, clear communication will play a pivotal part in ensuring that the patients experience in imaging is as good as possible, given the circumstances. From a practical viewpoint, all imaging should be conducted by manipulating the imaging equipment as opposed to patient position. This may require alternative views including horizontal beam lateral images and axial images. Radiographers in departments with DR systems may occasionally find CR receptors more useful in adapting technique for trauma, particularly where there is not a choice of sizes of DR detector (Hayre, Blackman, & Eyden, 2016). Where adaptation requires angulation of the central ray across the body of the patient, grids should not be utilized to avoid grid cut off.

If patients arrive in the department in inappropriate clothing or jewelry, the radiographer needs to make a decision whether it is appropriate to remove clothing in order to proceed with imaging. This decision should be discussed with the patient first, and where there is any risk to patient safety or risk of exacerbating injury, clothing should be left. Where possible, artifacts should be avoided in trauma imaging but it is important that the radiographer does not remove any equipment which is critical to patient safety, such as vacuum splints or spinal immobilization collar and blocks. The vast majority of emergency medicine accessories such as those mentioned are radiolucent to a degree and the risk of removal is not clinically justified.

SPINAL IMAGING IN EMERGENCY SETTINGS

As noted previously, a significant amount of the traditional general imaging workload has been transferred to some degree toward CT. There is perhaps no clearer example of this than the shift of approach in managing cervical spine trauma. NICE provides clear guidelines on the use of CT where any of the following are found:

- GCS of less than 13
- Intubation required
- Head injury or other additional trauma requiring scanning
- Patient aged 65 or over
- Focal peripheral neurological deficit
- Paresthesia in upper or lower limbs
- Suspicious or abnormal plain film findings
- Inadequate conventional imaging

(NICE, 2016)

Furthermore, NICE is clear that where imaging is appropriate for patients following cervical spine trauma, a three-view cervical spine series is appropriate. NICE makes no reference to the use of either the 'swimmers view' or oblique views to visualize the C7-T1 junction. It is, therefore, reasonable to suggest that where it is not possible to visualize the C7-T1 junction on a cross table lateral view, the next choice of imaging should be CT, in accordance with the NICE guidelines given the reference above to inadequate conventional imaging.

The radiographer performing imaging should first give significant consideration as to appropriateness of the use of a 'swimmers' view to visualize C7-T1 where this remains a departmental protocol due to the risks associated with the movement required to produce the image. The swimmers view also requires significantly increased exposure factors and therefore imparted dose in comparison to the lateral view. This is especially the case in imaging of larger patients who generally present the greatest difficulty in visualizing the cervicothoracic junction.

In any instance of cervical spine imaging following trauma, the lateral view of the cervical spine should always be the first image performed and the radiographer should consider the value of completing anteroposterior (AP) and 'Peg' (C1/C2) views, given that the patient should be referred on for CT imaging. It is important that the radiographer communicates with the medical team effectively where conventional imaging has not been successful in order to ensure the patient receives prompt CT imaging as appropriate.

Conventional imaging remains the first line of investigation for suspected injury to the thoracic and lumbar spine. AP and lateral views are generally sufficient for imaging of both the thoracic and lumbar spines. Lateral views should be performed with a horizontal beam with the patient remaining supine on the trolley. It is important that when producing these images the radiographers recognize the inevitable air gap between the patient and imaging receptor and exposure parameters are considered as a result.

RADIATION DOSE REDUCTION IN TRAUMA IMAGING

Trauma and emergency imaging can present many difficulties for a radiographer from a technical perspective. As patients in emergency imaging and trauma are rarely conventionally compliant, adaptation of technique, often accompanied by an adaptation to exposure factors, is an important skill for all radiographers. It is important in any situation that regardless of the potential pathology under examination, dose is kept as low as reasonably practicable.

As noted above, the use of grids in trauma imaging should be carefully considered, especially where significant adaptation of technique is required. Unless the radiographer can be certain that the central ray is perpendicular to the image receptor (and by extension the grid), a grid should not be used in order to avoid grid cut off artifact resulting in the need for repeat imaging and associated additional radiation dose to the patient. Where the vertical bucky is utilized, grids are unlikely to present an issue, provided that the imaging unit is locked into the correct position prior to exposure.

Often, in imaging requiring adaptation of technique, a larger source to image distance is required to compensate for an air gap, which is unavoidable due to patient position on the trolley. It is important that where a pre-set exposure is utilized, the radiographer is aware of the implications of increased distance on radiation intensity due to the inverse square law, and that exposure factors are increased appropriately. It is also worth considering that a grid may not be required where an air gap is present that would not normally occur in positioning with a standard technique.

Another method for reducing technique in trauma imaging, particularly in imaging of the pelvis, is to utilize the anode heel effect to the patient's advantage. Research has shown that in the case of male patients, gonad dose as a result of pelvis imaging can be significantly reduced by orienting the anode of the X-ray tube toward the patient's feet (Mraity, England, & Hoog, 2017). This is a simple but effective way to reduce dose without risking significant impact on image quality as a result of changing exposure factors for imaging.

ROLE OF THE RADIOGRAPHER IN IMAGE INTERPRETATION

While advanced practice radiographers will take responsibility for clinical reporting in the emergency setting (where there is a system in place to allow them to do so), there are still responsibilities for image interpretation by all radiographers. Most emergency imaging services will operate some form of radiographer abnormality detection scheme (RADS) to support decision-making by clinicians.

The 'red dot' system is perhaps the most common system across emergency departments in the United Kingdom (UK), with over 90% of departments operating the scheme (Snaith & Hardy, 2008). Within a red dot system, there is an expectation that radiographers will indicate a suspected abnormality, usually with a simple annotation on the image itself, or with a note added to the radiology information system event record. Red dot is a nonspecific form of alert, indicating that the radiographer believes there is an abnormality but providing no additional or clarifying information to the reviewing clinician.

Initial comment (or radiographer comment) is another type of RADS, but is not as widespread as the red dot system. Where initial comment is in use, radiographers are expected to provide a written record of their interpretation of the image produced. This has additional benefits in comparison to 'red dot' as referrers receive additional information, leading to fewer ambiguities. It is expected, for governance purposes, that any comment is written and recorded appropriately. It is important for referrers to be aware that an initial comment does not constitute a formal clinical report.

Regardless of which form of RADS is employed, it is important that appropriate governance procedures are adhered to. First, it should be identified if the RADS is voluntary or mandatory for radiographers in the department. Secondly, a method of ensure radiographers are appropriately educated to take part in the RADS is in place, or alternatively that radiographers are maintaining a suitable level of continuing professional development to take part effectively. Finally, an audit cycle should be in place to ensure the RADS is effective and useful.

GENERAL RADIOGRAPHY IN MASS CASUALTY INCIDENTS

A mass casualty incident (MCI) refers to any incident resulting in a higher number of casualties than reasonably expected. They may be of an intentionally violent nature, such as planned explosion or mass stabbing, or because of an incident such as a multivehicle road traffic collision, or such a collision involving a vehicle such as a bus leading to injuries to multiple people. It is important all radiographers are aware of their protocol for major incidents and mass casualty incidents, and these will often be found within the area of the department responsible for imaging for the ED. National guidance is also available from NHS England (2018).

The first steps to be taken in an MCI are to ensure the department is prepared to receive casualties. All nonurgent patients should be asked to leave the department to ensure rooms are freely available. Equipment should be on and ready for use. During preparation for receiving patients, it is also important that image intensifiers and mobile units are readily available in the areas they will be required to reduce delays to imaging. The department should also review staffing, and begin to call in additional radiographers to support throughout the major incident. Radiographers should be aware that significantly more staff than normal would be working in areas such as resuscitation where mobile imaging is likely to be required. Therefore, as part of preparation, it must be ensured that sufficient lead equivalent shielding is available to staff in those areas so that imaging is not delayed upon patient arrival.

In terms of organization, radiographers should be paired up where possible to allow one to position the patient and equipment, and the other to work at the computer in the room and manage the administration and record keeping. Staff should be assigned to specific areas and await patients in those areas. Staff may be given bibs or cards to indicate their role in the department to other members of staff. The organization of the department should be overseen by an experienced or lead radiographer acting as the imaging controller, who is easily identifiable to members of the team and to wider hospital staff.

During major incidents, casualties will be prioritized based on severity and will be directed to hospitals appropriate to their status. Major trauma centers will accept the most serious patients – designated at the scene as P1. Less-seriously injured patients, designated as P2, will be sent to trauma units, which do not necessarily have major trauma services. A report will be generated to indicate how many casualties each hospital should expect, and a member of the imaging team should be nominated to act as a liaison with the central controller and keep the imaging department updated and vice versa.

When P1 casualties arrive, some will be directed immediately CT for a full body scan. Others may be deemed too unstable for CT and may require mobile chest or pelvis imaging in resuscitation while the medical teams attempt to stabilize the patient. P2 patients may not require full body CT imaging and may have appropriate conventional imaging of affected areas of the body. During a major incident, requests for imaging should adhere to the standard legislation, but will not include real patient identifying data. Instead, patients will be assigned identifying numbers, which will be aligned with their records following the major incident. Radiographers should be aware that patients who have experienced an MCI may be in shock or be experiencing acute emotional responses and distress, and the radiographers should be prepared to manage this appropriately.

SUMMARY

Overall, this chapter has identified that in both critical care and trauma settings, strong interprofessional communication and teamwork are vital to achieving quality imaging and therefore better patient outcomes. It is important to remember that the radiographer, in these settings, is part of the wider team of professionals involved in the care of the patient and should work in a collaborative way accordingly. From a technical standpoint, imaging in both critical care and trauma is not without its challenges and these can vary depending on the patient and situation(s). It is the role of the radiographer to have awareness of patient safety in these settings and to carry out their work thoughtfully and deliberately. In trauma settings, where time is often a more significant pressure than in critical care, the safety of the patient should not be sacrificed in an effort to 'speed up' imaging. The reduction of dose – as part of patient safety – and the strong fundamentals in radiation protection, in addition to the illustrative ideas presented in the preceding sections, should not be forgotten in imaging in these areas.

REFERENCES

Brock, D., Abu-Rish, E., Chiu, C.-R., Wilson, S., Vorvick, L., Blondon, K., et al. (2013). Interprofessional education in team communication: Working together to improve patient safety. *BMJ Quality and Safety, 22*, 414–423.

Hayre, C. M., Blackman, S., & Eyden, A. (2016). Do general radiographic examinations resemble a person-centred environment. *Radiography, 22*(4), e245–e251.

Henschke, C. I., Yankelvitz, D. F., Wand, A., Davis, S. D., & Shiau, M. (1996). Accuracy and efficacy of chest radiography in the intensive care unit. *Intensive Care Radiology, 34*, 21–31.

Mraity, H. A., England, A., & Hoog, P. (2017). Gonad dose in AP pelvis radiography: Impact of anode heel orientation. *Radiography, 23*(1), 14–18.

NHS England. (2018). *Clinical guidelines for major incidents and mass casualty events.* London: NHS England.

NHS Improvement. (2016). *Resource set: Initial placement checks for nasograstric and orogastric tubes.* London: NHS Improvement.

NICE. (2016, February 17). *Spinal injury: Assessment and initial management.* Retrieved from www.nice.org.uk/guidance/ng41/resources/spinal-injury-assessment-and-ini tial-management-1837447790533

Rogers, F. B., Rittenhouse, K. J., & Gross, B. W. (2015). The golden hour in trauma: Dogma or medical folklore? *Injury: International Journal of the Care of the Injured, 46*(4), 525–527.

Snaith, B., & Hardy, M. (2008). Radiographer abnormality detection schemes in the trauma environment – An assessment of current practice. *Radiography, 14*(4), 277–281.

Tavangar, H., Shahriary-Kalantary, M., Salimi, T., Jarahzadeh, M., & Sarebanhassanabadi, M. (2015). Effect of family members' voice on level of consciousness of comatose patients admitted to the intensive care unit: A single-blind randomized controlled trial. *Advanced Biomedical Research, 4*(106).

10 Pediatric Imaging in General Radiography
Reflections of Practice

Allen Corrall and Joanna Fairhurst

INTRODUCTION

Children are not small adults, and radiographer approach, departmental protocols, and exposure factors in line with IR(ME)R to optimize patient dose to be as low as reasonably practicable (ALARP) should reflect this. A knowledge of this difference can prevent unnecessary imaging, for example, scaphoid views in the child under five, or guide appropriate imaging for referrals from specialist pediatric clinics. This chapter is not a positioning atlas; it aims instead to give an insight into the underlying reasons for why a child may need imaging and how to maximize the quality of your images while minimizing anxiety and distress in the child and their parents. Following a brief discussion of general principles in pediatric radiography, we will explore specific areas of practice and presentations to illustrate how these principles are applied.

PREPARING TO IMAGE A CHILD

Environment and preparation is a key essential when imaging children. Like the pit stop crew in a race, everything is prepared before the car arrives, but the objective is high image quality and the least distressing experience possible, not the speed at which it is done. This is easy to forget in a busy department, but put yourself in the child's shoes: they may be in a lot of pain and this may be their first experience of a hospital. Think how scary the room looks from their eye level, not forgetting the accompanying parent who is worried about their child's injury or illness. Every patient is an individual, not something on a production line that needs X-rays to then be passed on to the next department. While you might not get any response, interaction with the child is paramount from the moment you call their name. Most children will present with an opening to talk to them about, from a character on their clothes to the teddy bear that has clearly gone everywhere with them from the day they were born.

Excluding pregnancy in the female of child-bearing age takes on a new level of complexity in the pediatric patient. If handled insensitively, this can be embarrassing for the patient and unintentionally imply that a young child is sexually active, leading to complaints due to a poor approach to IR(ME)R. The author working with colleagues within a specialized pediatric department developed a simple pregnancy exclusion form that was well received in a Care Quality Commission IR(ME)R inspection. This explains that the radiographer is legally obliged to establish pregnancy status regardless of age or circumstances, irrespective of any physical or mental impairment, and goes on to ask if the child has started their periods and if so, their pregnancy status. The radiographer, if appropriate, then has the opportunity to apply the 10- or 28-day rule by ascertaining the date of their last menstrual period if the patient is unsure of their pregnancy status, and depending on the answer, only needing to ask if necessary the question, "can you confirm that you are not pregnant?" This has made the approach simple and consistent. The signed form is then scanned onto the patient's exam attendance.

COMMUNICATION

Extensive and adaptive communication skills are an essential quality required in a pediatric radiographer. For example a brief introduction of who you are to the parents of the new-born requiring an X-ray on the neonatal unit who also needs to recognize the parents' situation: they have just had their lives turned upside down and are in a daze struggling to comprehend the sudden onslaught of incoming information and flurry of activity around their new-born baby, who they haven't even had the chance to hold. Likewise dealing with the distressed child in the emergency department – in pain and frightened by the unfamiliar environment around them – requires patience and empathy. Inter-professional communication is vital to ensure your patient receives the highest standard of care and everyone involved in the process has the relevant information to minimize delays in treatment, especially if imaging reveals anything that needs to be addressed quickly.

PARENTS AND CARERS

Tailoring your language is important: use of technical terminology should be avoided. Instead provide a careful explanation of what the examination entails and if assistance from them is required, how you will need them to support you, with practical demonstration where required to avoid misinterpretation of what you have asked for. Children are very astute at picking up on anxieties whether it is their parents or your own. They look to their parents for reassurance in an unfamiliar situation and by effectively putting an anxious parent at ease you will help facilitate their encouragement alongside yours to complete the examination. In the event of the child becoming inconsolable a successful outcome can still be achieved if the parent has a clear understanding of what they need to do to assist as your objective shifts from avoiding distress to effectively minimizing the time the patient is in distress.

YOUR PATIENT

While nothing will guarantee a scream- and tear-free child, steps to avoid distress will help. Gentle play and interaction during identification checks help to form trust between you and the baby. Babies 'talk' with their eyes (Valman, 2010) and only faces and objects close by are in focus. A smiling face and using gentle speech while interacting with their parents while still looking at the infant will distract them as you gently get them into position. The familiar voice and touch of a parent is important and you must ensure that the baby can see and hear them throughout to help prevent distress.

As the infant's social and language skills develop, the importance of your effective interaction increases. From the age of six months, shyness or anxiety can manifest if approached abruptly or you get too close too soon, at nine months the mere approach of a stranger can present a challenge, so interacting with parents is vital (Sharma & Cockerill, 2014). Simple games such as peek-a-boo played alongside their parent reassures the infant as they look to their familiar adult for assurance in new situations, and they understand reference to mummy/daddy and look for them when asked where they are. (Sharma & Cockerill, 2014) You can use this to your advantage by using as a distraction technique, while one parent supports the child in position, the other can distract from behind the control panel.

At one year of age picture books can be used more effectively than at a younger age as the attention span increases, or (if available and it does not interfere with the examination) a noisy toy. A tuneful parent singing their favorite nursery rhyme is also a useful distraction tool! (Sharma & Cockerill, 2014). At 15 months the infant becomes much more physically restless, which could mean your window of best cooperation may be short lived, so the preparation of both room equipment and the parent's awareness of what is required will help you maximize the opportunity to complete the examination before restlessness turns to frustration and tears from being held still. Not rushing into the examination straight away and playing give and take games with a small toy, and teasing by offering and taking away may help distract from the new surroundings and instill some trust while you gently transition from playing to getting positioned for the X-ray (Sharma & Cockerill, 2014). At 18 months the infant begins to understand intentions from others expressions and actions so make positioning part of a game, if pads are going to be used allowing the infant to play with them first helps them realize they are not going to hurt them and may reduce their aversion to having them touching them.

From around two years of age the infant understands when communication is addressed to them and will join in with action songs and their favorite nursery rhymes and is able to point to named body parts so understand references to where you are going to X-ray (Sharma & Cockerill, 2014). Frustration tantrums are usually easily overcome by distracting, but they will defend their possessions with determination (Sharma & Cockerill, 2014); so be careful not to inadvertently make things worse because they think you are trying to steal their toy!

At 2½ the infant is even more active and much more restless and resists restraint and once distressed is much more difficult to distract until removed from the situation that caused the distress. A well-informed parent is key here as the infant looks to them for reassurance.

The three-year-old toddler if not overcome by shyness many give you their full name and sometimes their age. They will have a favorite story or nursery rhyme you can get the parent to recite. Although they do not understand the concept of time yet, they can count a little (Sharma & Cockerill, 2014), so use of slowly counting up for how long to stay still for, aiming to finish before getting to the agreed number can be a useful tool. The four-year-old starts to understand the concept of before, now, and after so you can begin to manage their expectations for the progress of the examination. The five-year-old may be able to also tell you their date of birth along with their full name and will probably have a favorite story or joke they will enjoy telling you (Sharma & Cockerill, 2014). They will have an understanding of time and event sequences which you can again use to manage their understanding of how long the examination is going to take.

With every child it is absolutely essential you remain on or below their eye level (Valman, 2010), so for the child carried into the X-ray room and then placed down on the floor while their parent gets ready to help you, you must follow them down and crouch to remain on their level so you are not towering above them. You can use this opportunity to initiate simple conversations with them by asking things like 'What did you have for breakfast/lunch before you came here?' or 'Did you come in mummy/daddy's car or on a bus?; (Valman, 2010).

IMAGING THE NEONATE

Survival rates of the extremely premature neonate have increased as knowledge and technology have developed. The World Health Organisation (WHO) defines prematurity as 'babies born alive before 37 weeks of pregnancy are completed'. There are sub-categories describing preterm birth, based on gestational age (Figure 10.1).

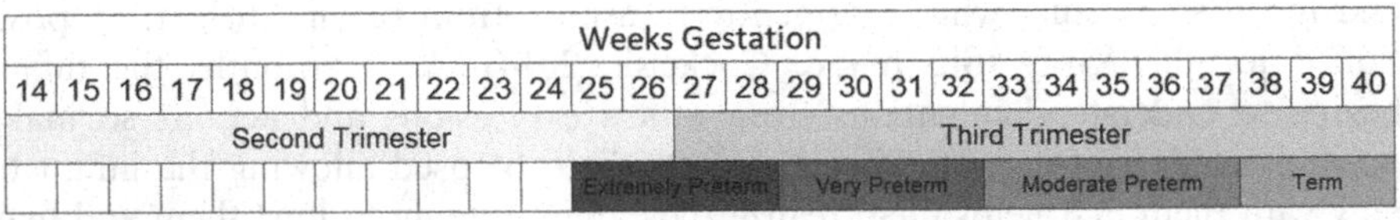

Weeks Gestation																										
14	15	16	17	18	19	20	21	22	23	24	25	26	27	28	29	30	31	32	33	34	35	36	37	38	39	40
Second Trimester													Third Trimester													
											Extremely Preterm		Very Preterm			Moderate Preterm				Term						

FIGURE 10.1   Prematurity classification.

Throughout their stay in the neonatal unit the neonate can undergo many X-rays. The risk factor of radiation induced cancer in later life is therefore increased. The neonate is rapidly producing blood cells and depending on their gestational age the production site changes. The liver takes over from the yolk sac and is the primary producer of blood cells from around 15 weeks gestation declining steadily until 40 weeks, production in the bone marrow starts around 15 weeks and overtakes the liver in production around 22–23 weeks.

Protecting these areas with additional lead on the top of the incubator or using a 'Table top' in the open incubator can be used to further collimate the X-ray beam helps to minimize dose to these radiosensitive areas (Figure 10.2).

FIGURE10.2   'Table top' with additional lead.

THE NEONATAL CHEST

Whilst in utero the fetal blood flow bypasses the fluid filled lungs via the ductus arteriosus, which connects the pulmonary artery to the aorta, and the foramen ovale, which allows blood to flow from the right atrium to the left.

Newborns only have around 50–70 million alveoli, compared to an adult with around 300 million. For the neonate born at term, hormones triggered by the onset of labor cause some of the fluid in the alveoli to be reabsorbed and most of the remaining fluid is squeezed out during the journey through the birth canal. The remaining fluid continues to be reabsorbed and is also expelled through normal respiration (South & Isaacs, 2012) which is why unless clinically necessary, a chest X-ray should be delayed until four hours of age to allow this remaining fluid to be lost as it can be mistaken for pathology, subjecting the neonate to unnecessary treatment. For the baby born via caesarean section the volume of fluid in the lungs is higher as the natural squeezing of the chest does not occur and they are more likely to suffer from transient tachypnoea of the newborn (TTN). It is good practice to annotate a chest X-ray image with the time of birth if being imaged on the date of birth to allow the reporting radiologist to take into account the possibility of remaining lung fluid being mistaken for a pneumonia.

As the neonate's lungs naturally recoil and they start to take their first few breaths, blood fills the lung capillaries and the resistance within the pulmonary vessels falls, the airways fill with air and gaseous exchange starts within the alveoli. As the umbilical cord is clamped, the peripheral blood pressure increases causing a pressure differential within the heart closing the foramen ovale: this closes permanently with the fall in maternal hormone levels that had kept this patent in the uterus (South & Isaacs, 2012). The ductus arteriosus usually closes within a few days of birth. For the term neonate with mature lungs, the recoil and entry of air into the alveoli causes surfactant to be released. This is a substance that reduces the surface tension of the alveoli and prevents them collapsing during exhalation. In the preterm infant with immature lungs this process does not occur. If there is a high risk of premature labor the mother may be given steroids from 23 weeks gestation to aid in surfactant maturation, and the preterm neonate given surfactant soon after birth. This is one occasion when the 'four hour rule' for a chest X-ray is not applicable, as the exogenous surfactant has to be delivered via an appropriately sited endotracheal tube.

Hypoxic-ischaemic encephalopathy (HIE) is the result of a hypoxic-ischemic insult which occurs when blood and oxygen flow to the fetus is interrupted either in the uterus or during birth (Tasker, McClure, & Acer, 2013). This can occur for many reasons, for example placental abruption or problems involving the umbilical cord. If HIE is recognized fast enough (within 6 hours) the infant's body temperature is reduced to 33–34°C and this induced hypothermia is maintained for 72 hours before the neonate is gradually rewarmed (Tasker, McClure, & Acer, 2013). Induced hypothermia has improved the prognosis which varies depending on the duration of the hypoxic-ischemic insult and the speed at which it was recognized.

MECHANICAL VENTILATION

In their first few breaths the neonate establishes their total lung capacity, residual volume, and tidal volumes (the amount of air inhaled/exhaled at rest). The air within the airways (nose, trachea, and bronchiolar structures not actively involved in gaseous exchange) is known as dead space.

Mechanical ventilation carries inherent risks and in the fragile neonatal lung can lead to ventilation induced injury. Misplaced tubes are a common cause of collapse, which can be mistaken for consolidation particularly if the right upper lobe is involved.

Common ventilation methods used in a neonatal setting are as follows:

Intermittent positive-pressure ventilation (IPPV)
Positive end-expiratory pressure (PEEP)
Constant positive air pressure (CPAP)
Bi-level positive air pressure (BiPAP)
High-frequency oscillatory ventilation (HFOV)

(Tasker, McClure, & Acer, 2013)

IPPV is delivered via the endo-tracheal tube (ETT) and allows the neonate to breathe naturally. With PEEP, the oxygen pressure is raised at the end of the expiratory phase to help prevent alveolar collapse while the neonate passively exhales. CPAP, as the name implies, utilizes a constant oxygen flow delivered at low pressure as the neonate breaths spontaneously, the CPAP keeping the alveoli inflated to prevent them collapsing during the breathing cycle. BiPAP alternates between two pressures to assist with exhalation. HFOV maintains a set pressure with rapid pressure variation around this level at very high frequency (up to 900 times a minute) which creates very small tidal volumes, often less than the dead space volume.

A complication of mechanical ventilation is pulmonary interstitial emphysema (PIE) (Tasker, McClure, & Acer, 2013). This is where the alveoli are ruptured and air leaks into the surrounding supportive tissues of the lungs. The lungs appear hyper inflated on chest X-ray with a 'honeycomb' pattern or bullae: large dilated air-filled spaces.

The left-hand image (Figure 10.3) was taken not long after premature birth and the infant is already being ventilated, the one on the right obtained 10 days later shows extensive bullae.

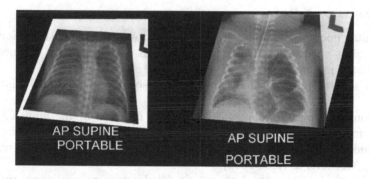

FIGURE 10.3 Ventilator induced injury.

RESPIRATORY PROBLEMS PRESENTING SOON AFTER BIRTH

These include respiratory infection, most commonly acquired from the mother's genital tract (congenital pneumonia), which usually becomes apparent within the first 24 hours of life (Tasker, McClure, & Acer, 2013).

If the neonate becomes distressed during childbirth they can pass meconium, a substance within their intestines composed of ingested materials while in the uterus (meconium aspiration) (Tasker, McClure, & Acer, 2013). This blocks the airways, causing focal collapse if the obstruction is complete and focal hyperinflation if the airway is partially occluded, and results in a chemical pneumonitis.

TTN usually presents within four hours of birth is caused by delayed clearing of lung fluid and is more common in the child born by Caesarean section (Tasker, McClure, & Acer, 2013).

The preterm neonate is at particular risk of spontaneous pneumothorax due to the under-developed lung tissue, especially fragile as alveolar lining is only one cell thick to enable gaseous exchange across the membrane. The ventilated neonate is also at risk of a tension pneumothorax, an acute life-threatening condition exacerbated by the constant positive air pressure used in certain ventilation setups. Gas enters the space between the pleura and builds up as the positive ventilator pressure prevents backflow. The pleural gas collection pushes the trachea and mediastinum to the opposite side of the air leak and compresses the venous great vessels leading to circulatory compromise. Presence of chest drains may not prevent or resolve the tension pneumothorax if the drain is not appropriately placed, or if pressure from the tension pneumothorax is compressing the lung tissue onto the drain, hence occluding it – one reason why a ventilated neonate may develop worsening saturation and their chest drains have stopped bubbling.

A congenital diaphragmatic hernia presents immediately with respiration difficulties due to a defect in the diaphragm allowing abdominal contents to herniate into the chest.

Appearances on X-Ray

Congenital pneumonia shows as patchy shadowing and consolidation, or as ground glass shadowing. This appearance can be mimicked if the neonate is imaged on a mattress.

TTN shows streaky perihilar changes and sometimes fluid in the horizontal fissure.

Meconium aspiration shows as focal over inflation of the lungs caused by the incompletely blocked airways alternating with patchy areas of collapse.

Pneumothorax show as air filled space but lacking any lung markings. (Tasker, McClure, & Acer, 2013).

Tension pneumothorax show as air filled space as per a pneumothorax with midline shift of the trachea and mediastinum.

In congenital diaphragmatic hernia X-ray may show loops of bowel within the chest.

THE NEONATAL ABDOMEN

Necrotising Enterocolitis (NEC)

The preterm neonate is more at risk of developing necrotising enterocolitis owing to the fragility of the immature intestine. Intestinal inflammation ranges from mucosal damage to complete necrosis and perforation. Proximal colon and terminal ilium are the most common sites affected. If free air is not visualized on a chest or abdominal X-ray, the neonate can be placed on their left side for five minutes (left lateral decubitus projection) which will show any free abdominal air outlined by the liver. Advanced NEC can lead to multi-organ failure. Treatment is two-fold, antibiotics to treat the infection and parenteral feeding (intravenous nutrition) while the intestines recover before reestablishing gastric based feeding either by breast or nasogastric (NG) means. It is notable that the incidence of NEC is reduced six-fold in breast-fed preterm neonates (Tasker, McClure, & Acer, 2013).

Bell Staging of NEC:

Stage I: Suspected NEC
Abdominal X-ray only shows bowel distension

Stage II: Definite NEC
Abdominal X-ray shows distension and bowel wall thickening, sometimes also shows intramural gas or a fixed bowel loop

Stage III: Advanced NEC
Abdominal X-ray possibly shows pneumoperitoneum
                                        (Tasker, McClure, & Acer, 2013)

If the neonate has failed to pass meconium within the first 24 hours of life, two common causes of abdominal obstruction are Meconium Ileus and Hirchsprung's Disease.

Meconium ileus is common in cystic fibrosis, which is an autosomal recessive genetic disorder with one symptom being thickened secretions. These affect both lungs and intestinal tract (Tasker, McClure, & Acer, 2013). Thicker and stickier meconium becomes inspissated causing an abdominal obstruction evidenced by dilated bowel loops above the point at which the meconium plug is located. A therapeutic enema utilizing water-soluble contrast agent such as dilute Gastrografin may help loosen the meconium and therefore hopefully avoid the need for surgical intervention (Tasker, McClure, & Acer, 2013). This procedure is considerably less risky than in previous decades since it was shown that the therapeutic effect relies on the presence of surfactant in the contrast agent rather than its hyperosmolarity.

Hirchsprung's disease is commonly associated with Trisomy 21 (Down syndrome): the neonate fails to pass meconium due to the partial or complete absence of innervated intestine. This leads to obstruction, as intestinal contents cannot be propelled through the area of denervated bowel. Abdominal X-rays show a distal abdominal obstruction. (Tasker, McClure, & Acer, 2013)

CHALLENGES IN THE NEONATAL ABDOMEN

Gastroschisis denotes bowel outside the abdomen which is not covered by the peritoneum and has thickened due to contact with amniotic fluid and exomphalos (omphalocele) is where the intestine and sometimes the liver herniates outside the abdominal cavity but a membranous covering is retained (Tasker, McClure, & Acer, 2013). Neither require an X-ray for diagnosis, but present challenges to the radiographer as they can be a source of artifact and require careful positioning to minimize the effect on imaging, even on chest X-rays, especially if an exomphalos is extensive.

MALROTATION

Malrotation can occur if the bowel does not rotate and fix into the correct position within the abdomen during its development. The abnormally fixed intestine is at risk of twisting and this commonly presents with bile-stained vomiting. An abdominal X-ray is performed to exclude other conditions but a fluoroscopic contrast study will generally be required to assess the position of the duodeno-jejunal flexure (which is abnormally sited) and to exclude volvulus, a life-threatening complication when the gut twists and its blood supply is cut off.

INTUSSUSCEPTION

This is when the intestine telescopes in on itself. This is a serious condition and if undetected or untreated can cause the affected part of the bowel to die and can be further complicated by perforation, leading to sepsis and in the worst case scenario the death of the infant. Fast intervention is critical to maximize the chances of successfully reducing the intussusception with no further complications and full recovery. An ultrasound scan is performed to confirm the diagnosis, and if there is no evidence of perforation or peritonitis a reduction can be performed by an air or contrast enema done under fluoroscopic guidance. If this fails or there are contraindications to the procedure the infant is taken immediately to theatre to undergo a surgical reduction.

LINES AND CONED VIEWS

PERIPHERALLY INSERTED CENTRAL CATHETER (PICC)

A durable catheter that can be left in place for long-term treatment with the insertion point in either upper (one of the veins running through the antecubital fossa) or lower limbs (long saphenous vein). Due to placement in the peripheral venous system the path it can follow is variable and X-ray is required to confirm that the tip sits in the superior vena cava. These lines utilize the high blood volume in the vena cava to inherently reduce the irritant or osmotic effects of some medications that would irritate the smaller veins causing phlebitis. PICC lines can also be used for blood sampling.

CENTRAL VENOUS CATHETER (CVC)

As with PICC lines, these are used for long-term delivery of medication, but their entry point into the venous system is into one of the greater vessels such as the internal jugular or subclavian vein (Tasker, McClure, & Acer, 2013).

UMBILICAL ARTERIAL CATHETER (UAC)

These can be inserted up to 48 hours after birth and can be used for continuous blood gas monitoring, blood pressure measurement, infusions, and transfusions (Tasker, McClure, & Acer, 2013).

UMBILICAL VENOUS CATHETER (UVC)

These can be inserted up to five days after birth and can be used for emergency access during resuscitation, central venous pressure measurement, infusions, and transfusions and if peripheral access is difficult, especially in the preterm neonate (Tasker, McClure, & Acer, 2013).

CARDIAC CONDITIONS

ATRIAL SEPTAL DEFECT (ASD) (LEFT TO RIGHT SHUNT)

An atrial septal defect allows oxygenated blood in the left atrium to mix with deoxygenated blood in the right atrium (Tasker, McClure, & Acer, 2013). This is not to be confused with foramen ovale not closing correctly after birth, which is patent foramen ovale (PFO). The defect affects the part of the septum that is involved with the development of the mitral and tricuspid valves allowing oxygenated blood to flow from the left atrium to the right reducing the amount of oxygenated blood to be pumped around the body. This can lead to the heart enlarging as the high blood pressure from the left atrium causes the right side of the heart to dilate to accommodate the blood flowing through the defect. This type of defect is often seen in babies with Down syndrome.

VENTRICULAR SEPTAL DEFECT (VSD)

Most ventricular septal defects close spontaneously: large defects present clinically with breathlessness after the first few days of life due to heart failure (Tasker, McClure, & Acer, 2013). This is due to some of the blood at high pressure in the left ventricle flowing into the right ventricle and into the lungs. The heart then has to work harder to supply enough oxygenated blood to the body. If the defect is too large and causes heart failure, the loss in pumping efficiency leads to fluid building up in the lungs: in these circumstances an operation to close the hole will be required.

PERSISTENT DUCTUS ARTERIOSUS (PDA)

This is when the hole that allows lung bypass while the fetus is in the womb is still present one month after the date the baby was actually due (Tasker, McClure, & Acer, 2013). Some heart conditions such as the transposition of the great arteries require this to remain open and steroids are given to ensure this until the defect is corrected. If the defect does not close spontaneously and causes heath issues it can be closed surgically or percutaneously.

TETRALOGY OF FALLOT (TOF) (RIGHT TO LEFT SHUNT)

This can be an isolated anomaly or may occur in several syndromes, the most well-known being Down syndrome. The neonate becomes cyanosed due to the combination of cardiovascular defects and they can even lose consciousness when under stress as the heart cannot supply enough oxygenated blood to the body. It is caused by four anatomical anomalies (Tasker, McClure, & Acer, 2013): A large VSD;

1. Overriding Aorta – the aorta is positioned over the VSD not the left ventricle, allowing deoxygenated blood into the aorta to be pumped around the body;
2. Pulmonary stenosis – The pulmonary artery is narrowed, increasing the heart's workload in pumping blood to the lungs, leading to …
3. Right ventricular hypertrophy – Increase in heart size due to the extra work the heart needs to do to pump blood into the lungs
4. Treatment is usually undertaken in stages to allow the heart to recover. The stenosis is usually treated first and can be corrected by either balloon dilatation of the stenosis or an artery can be diverted to increase blood flow to the lungs. The ventricular defect is surgically repaired.

ERB'S PALSY

This is a birth related injury which can occur several ways, most commonly caused by shoulder dystocia where the baby becomes stuck in the birth canal as one of its shoulder lodges on the mother's pubic bone and during the associated assisted delivery the head (or trunk in breech birth) is pulled causing nerve damage as the neck is stretched (South & Isaacs, 2012). The neonate presents with upper limb paralysis on the affected side (sometimes both) and the X-ray request will often mention lack of tone in the affected arm. The role of X-ray is to exclude a fractured clavicle or humerus as an alternative cause for paresis. Nerve damage sequelae range from full recovery to a permanent degree of paralysis.

ORTHOPEDICS

Skeletal development begins in the third week of gestation with the term neonate having an impressive 300 bones (including epiphyses), development continues into early adulthood with the mature skeleton containing 206 bones.

In the fetus the notochord becomes the vertebral column, by week 4 the limb buds form, then the sacrum and then the facial bones. By weeks 5–8 the fetus takes on a recognizable shape.

Bone is a living, highly vascularized connective tissue and a reservoir for minerals and nutrients and is constantly being made by osteoblasts and broken down by osteoclasts.

TYPICAL PARTS OF A BONE

Bone is made up of several constituent parts as shown in Figure 10.4 below.

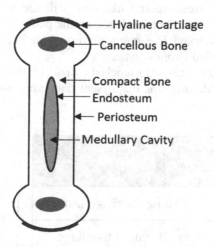

FIGURE 10.4   Typical parts of bone.

Hyaline cartilage – Coats ends of bones where a joint is formed

Endosteum – Inner aspect of the bone – lines the medullary cavity, plays an important part in the repair of fractures

Periosteum – Outer aspect of the bone – forms a protective outer-casing and point of attachment for muscles and tendons

Medullary cavity – Contains yellow bone marrow in adults

Ossification can be intramembranous or endochondral.

Intramembranous ossification starts early in fetal life and is the process particularly responsible for plates of bone such as the bones of the skull. Endochondral ossification (replacement of cartilage 'chondro' by bone 'ossify') also usually starts in the fetus by development of the cartilaginous template from mesenchymal cells (cells that can differentiate into osteoblasts and chondrocytes). Invasion of chondroblasts helps the template grow in length and width and by the end of the process the beginnings of the medullary cavity develop. A nutrient artery forms and pierces the perichondrium (the perichondrium becomes the periosteum), and the increased blood supply stimulates perichondrium to specialize into bone cells to form a collar of bone (the periosteum). The spongy bone of the diaphysis starts to form and is eventually replaced by compact bone.

As with any medical terminology the prefix gives important information about the developing part of the bone:

Epi – Before
Meta – After
Dia – Complete

Bone grows due to processes occurring at the epiphyseal plate as shown in
Figure 10.5, which can be divided into four zones:

1. Small scattered chondrocytes anchor the epiphyseal plate to the diaphysis
2. Large chrondrocytes (stacked like coins) divide to replace the dying
   chondrocytes on the diaphyseal side of the plate.
3. Chondrocytes stacked like coins mature with highly increased bone
   function – calcified chondrocytes.
4. Calcified chondrocytes – Osteoclasts dissolve the new cartilage and
   osteoblasts invade from the diaphyseal plate and ossify the template

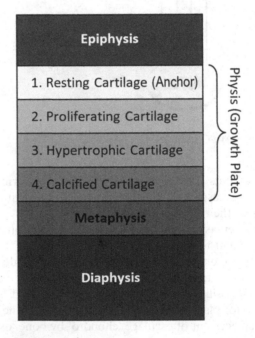

FIGURE 10.5   Growth plate cartilage layers.

Zones 2 and 3 are where bone growth occurs. As the bone reaches maturity
the epiphyseal plate gets narrower and the epiphysis and diaphysis eventually
fuse. If all the zones are damaged in a child, growth can be stunted.

In females the final growth plates fuse at around 18 years of age, males at
around 21 years old, leaving a visible dense epiphyseal line. The clavicle is the
last bone to stop growing.

FRACTURE PATTERNS IN THE GROWING SKELETON

SALTER-HARRIS FRACTURES

This classification system describes fractures involving the growth plate is involved in relation to the fracture (Figure 10.6).

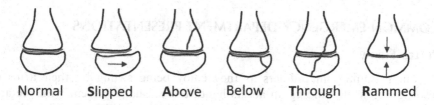

Normal   Slipped   Above   Below   Through   Rammed

FIGURE 10.6   Salter-Harris fracture classifications.

Type I: Slipped	Fracture of the physeal cartilage not involving either metaphysis or epiphysis
Type II: Above	The fracture extends along the growth plate and then into the metaphysis
Type III: Below	The fracture extends along the growth plate and then into the epiphysis
Type IV: Through	The fracture line passes from the metaphysis, through the physeal plate and into the epiphysis
Type V: Rammed	Axial compression of the physeal plate

BOWING FRACTURE

Due to the higher proportion of cartilaginous content, a child's bone is more pliable than an adult's and can bend to a higher degree before a fracture occurs. Occasionally the bone may remain deformed micro-fractures occur within the cancellous bone but the cortex remains intact.

GREENSTICK

As the bone bends, one cortex undergoes compression and remains intact or buckles but the opposite side fractures as its elastic limit is exceeded, breaking in the same fashion as the young branch of a tree does when bent, hence the name 'Greenstick'.

BUCKLE/TORUS

These occur from an axial compression such as a fall onto an outstretched hand, which compresses the cortex, either on one side or the complete circumference, without cortical breach.

COMPLETE

Both cortices are fractured. The inclination of the fracture line depends on the direction of the force causing the fracture. Transverse fractures usually occur due to a direct impact, such as the bumper of a car. Spiral (oblique) fractures occur when the bone is subjected to a twisting force, such as when a toddler is learning to walk and trips over their own feet and their body weight is transmitted through the leg they tripped over (Toddler's fracture) (South & Isaacs, 2012).

COMMON EMERGENCY DEPARTMENT PRESENTATIONS

PULLED ELBOW

A common injury in toddlers as they enjoy being swung by their arms or from falling while holding an adults hand, these usually present with the arm straight whereas fractures tend to present with the elbow flexed and held across the body. Usually reduced by fully flexing the arm and supination of the elbow, they can occasionally relocate as you position for a textbook AP elbow, relocating with a satisfying clunk (South & Isaacs, 2012) and the child immediately settling down as the subluxation is reduced.

SUPRACONDYLAR FRACTURES

This is common in the 4–10-year-old age group. This fracture puts the radial, ulnar, and median nerves at risk (South & Isaacs, 2012), and if the fracture is displaced or becomes displaced due to radiographer error in moving the patient and not adapting technique, damage to the brachial artery can also occur. Adaptive technique is vital both to prevent further injury and to minimize the patient's discomfort. A flat pad to support the forearm and elbow can be used to manoeuver the limb only moving the shoulder joint and can be left in place for both vertical and horizontal ray imaging, minimizing the patient's movement.

HIP IMAGING

The child born with a developmental dysplasia of the hip (DDH), those who develop Perthes disease and some cerebral palsy suffers are likely to undergo a lot of imaging so protection of their gonads is paramount. Direct fenestration using lead rubber with cut-outs placed directly on the patient works well but can slip, especially in the distressed child. Indirect fenestration in varying sizes as shown in Figure 10.7 that slot directly into the x-ray tube to apply the same restricted field of view without the issues of movement and infection control works much more effectively. This can also be used in the emergency department for the frog lateral hip images in cases of suspected slipped upper femoral epiphysis (SUFE).

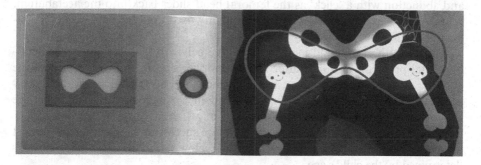

FIGURE 10.7 Indirect hip fenestration.

CEREBRAL PALSY

Cerebral palsy describes a range of disabilities resulting from an insult during brain development. Depending on when in it occurs during neuronal development and the extent of the insult, the child can be hemiplegic with normal intelligence but often suffers with visual deficits and epilepsy; diplegic, which predominantly affects the lower limbs with normal levels of intelligence, or quadriplegic with varying levels of developmental, visual and occasionally oromotor impairments, and poor trunk control requiring supportive wheelchairs. The Gross Motor Function Classification System (GMFCS) is used to show the mobilizing abilities of the child. This ranges from mild motor control problems causing clumsiness but otherwise fully mobile through to spastic quadriplegia with little or no muscular control requiring splints and orthoses.

Orthopedic interventions are commonly required in the hip, due to flexion and limited adduction, often with internal rotation, and for the ankle suffering from plantar flexion and either valgus or varus deformity of the foot.

Dyskinetic cerebral palsy is sometimes used to describe the resulting disability from HIE. The child suffers from involuntary movements such as dystonia (twisting and repetitive movements) or athetosis, (slow writhing movements of limbs) (South & Isaacs, 2012).

DEVELOPMENTAL DYSPLASIA OF THE HIP (DDH)

Previously known as congenital dislocation of the hip (CDH), this is the most common musculoskeletal abnormality in neonates, with the left hip being the most affected, thought to be due to fetal position in the uterus. A familial history increases risk and it is more common in firstborns and females, with a breech birth being the highest risk factor of all. It is less common in communities where the mother carries their baby in a sling facing them with the legs widely abducted. Normal Barlow and Otolani tests (South & Isaacs, 2012) performed at birth do not preclude DDH. The Barlow test looks for hips that can be dislocated under light pressure, but a dislocating hip can resolve spontaneously in the first few weeks of life so a positive result is not a clear indicator that DDH will be an issue. The Otolani test looks for a hip that is dislocated and relocates on flexion and abduction with a 'click' as the femoral head slides back into the acetabulum. Ultrasound is usually the first imaging modality used until the femoral head ossification centers become reliably visible.

The initial X-ray should include the whole pelvis to look for any other developmental problems and thereafter fenestrated views are employed to protect the gonads. A dysplastic hip demonstrates an increased acetabular index, a shallow acetabulum and a hypoplastic femoral head in a superolateral position as the shallow acetabulum allows it to sublux.

The earlier intervention takes place, the more successful the outcome, with most hips going on to develop normally. Table 10.1 below shows the treatment pathways determined by the child's age.

TABLE 10.1

Hip dysplasia treatment

Age	Treatment
< 6 months	Pavlik Harness (Figure 10.8)
6–18 months	Manipulation and closed reduction + hip spica cast *or* Open reduction + hip spica cast
18–24mths	Open reduction ± pelvic osteotomy + hip spica cast
2–6yrs	Open reduction + hip spica cast

Table adapted from data in Practical Paediatrics (7th Ed.)

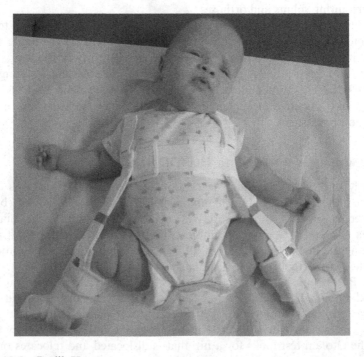

FIGURE 10.8 Pavlik Harness.

There are many systems of classifying and measuring hip dysplasia but all DDH fall into categories outlined below:

Stable
Subluxable
Dislocatable
Dislocated – Reducible
Dislocated – Irreducible
Teratological

(South & Isaacs, 2012)

PERTHES' DISEASE

Idiopathic in nature and four times more common in males. Caused by avascular necrosis of the femoral head, the disease is self-limiting but needs careful monitoring to ensure that during the necrotic stage the femoral head does not dislocate. AP and lateral views should be taken on initial presentation and coned AP views throughout monitoring (Tasker, McClure, & Acer, 2013).

Pathological Stages:

Avascular
The femoral head appears sclerotic with no or minimal loss of epiphyseal height
Fragmentation
Fissures appear, resulting in fragmentation of the femoral head with loss of epiphyseal height
Remodeling
New bone begins to form and the femoral head begins to remodel
Healed
The femoral head no longer appears avascularized on X-ray

SLIPPED UPPER FEMORAL EPIPHYSIS (SUFE)

A Salter-Harris Type I fracture. The upper femoral epiphysis slips on the metaphysis through the hypertrophic cartilage of the growth plate (Tasker, McClure, & Acer, 2013). Usually prevalent in 10–15 year-olds presenting with groin pain on the affected side and sometimes pain in the same knee (South & Isaacs, 2012). Full AP pelvis to exclude any other cause and coned frog laterals are required to assess the epiphyseal position.

SUFE Classifications:

Pre-slip
Acute slip
Acute on chronic
Chronic

(Tasker, McClure, & Acer, 2013)

Immediate surgical intervention is required in order to prevent the slip progressing and affecting growth. Follow up may include full leg-length imaging to assess any leg length discrepancy that develops.

IRRITABLE HIP

This condition of uncertain etiology has a clinical presentation similar to Perthes disease, or septic arthritis. AP pelvis and coned frog laterals should be done to assess for a possible SUFE or Perthes disease, but normal radiographs cannot exclude septic arthritis and joint aspiration may be necessary.

THE LIMPING CHILD

With no history of (witnessed) trauma, the diagnostic pathway will often commence with an X-ray of the presumed affected joint. However, if a knee appears to be the source of pain, hip X-rays may be indicated as pain can be referred to the knee, especially with a chronic SUFE. With young patients accurate pain localization can be difficult so the clinician will begin by excluding common conditions.

SEPTIC ARTHRITIS

This is most frequently a primary infection of the joint space but can develop into osteomyelitis as the infection spreads, especially in neonates (Tasker, McClure, & Acer, 2013). Initial X-rays of the joint may appear normal, but as the infection progresses the joint space may be widened as effusion within the joint increases. Later changes that may become apparent are subluxation or dislocation of the joint, narrowing of the joint space, with destructive changes in adjacent bone seen in later stages of infection.

OSTEOMYELITIS

Usually affects the metaphyseal regions (Tasker, McClure, & Acer, 2013) as the blood flow is plentiful but slow through the growth plate, allowing blood-borne bacteria to settle as this region is relatively protected from the body's immune response (South & Isaacs, 2012).

X-rays may not show any bone abnormality in the early stages, with soft tissue edema the only indication of infection. Late stages reveal reduction in metaphyseal density and destructive changes.

OSGOOD-SCHLATTER'S DISEASE

This is not a disease as such but an inflammation of the tibial apophysis: rapid growth at puberty can lead to the tendon between the patella and the fixation point at the tuberosity on the anterior aspect of the tibia to pull against the growth plate, causing inflammation. This is made worse during strenuous activity and treatment is usually resting the joint and stopping activities when it becomes painful. Although a clinical diagnosis, X-rays are frequently requested to exclude other possible diagnoses.

Sever's Disease

As with Osgood-Schlatter's disease this is inflammation of the insertion site of the Achilles tendon at the os calcis (South & Isaacs, 2012). If the only site of pain is the posterior aspect of the heel, only coned lateral calcaneum views are required.

Genu Valgum (Knock Knees)

When the child stands with their knees together, the medial malleoli do not touch. Full leg length X-ray imaging is required and surgical treatment using reversible epiphysiodesis (8-plates) are used to correct the alignment. The 8-plate restricts the growth of one side of the long bone and when the deformity is corrected the 8-plate is removed to prevent over-correction.

Femoral Anteversion

The whole leg is medially rotated from the hip and the patellae and feet look toward each other. This can self-correct as the child grows, but if it doesn't it can be surgically corrected by means of a varus derotation osteotomy. This is sometimes also combined with a Dega osteotomy to deepen a shallow acetabulum often seen in non-mobile cerebral palsy children.

ACKNOWLEDGEMENT

Thank you to my co-author Consultant Paediatric Radiologist, Dr. Joanna Fairhurst both for her professional support in my post as a Specialist Paediatric Radiographer and advice, input, proof reading, and assistance with this chapter.

REFERENCES

Carver, E., & Carver, B. (2006). *Medical Imaging Techniques, Reflection & Evaluation*. London: Churchill Livingstone.
Fairhurst, J. J. (n.d.). Emergency Paediatric Radiology.
Jacoby, J., & Ayer, R. (2009). *Frameworks for Radiology Reporting*. London: Royal Society of Medicine Press Limited.

Maconochie, I., Bingham, B., & Skellett, S. (2019, March 28). *Paediatric Basic Life Support*. Resuscitation Council (UK). Retrieved from: www.resus.org.uk/resuscitation-guidelines/paediatric-basic-life-support/#process.

Perkins, G., Colquhoun, M., Deakin, C., Handley, A., Smith, C., & Smyth, M. (2019, March 28). *Adult Basic Life Support and Automated External Defibrillation*. Resuscitation Council (UK). Retrieved from: www.resus.org.uk/resuscitation-guidelines/adult-basic-life-support-and-automated-external-defibrillation/.

Rudolf, M., Lee, T., & Levene, M. I. (2011). *Paediatrics and Child Health* (3rd ed.). Chichester: Wiley-Blackwell.

Sharma, A., & Cockerill, H. (2014). *Mary Sheridan's from Birth to Five Years: Children's Developmental Progress* (4th ed.). Abingdon: Routledge.

South, M., & Isaacs, D. (2012). *Practical Paediatrics* (7th ed.). Churchill Livingstone/Elsevier.

Staheli, L. T. (2003). *Fundamentals of Pediatric Orthopedics* (3rd ed.). Philadelphia: Lippincott Williams & Wilkins.

Tasker, R. C., McClure, R. J., & Acer, C. L. (2013). *Oxford Handbook of Paediatrics* (2nd ed.). Oxford: Oxford University Press.

Valman, B. (2010). *ABC of One to Seven* (5th ed.). Oxford: Wiley-Blackwell.

Wootton, R. (2008). *Radiation Protection of Patients (Postgraduate Medical Science)*. Cambridge: Cambridge University Press.

Section 4

Learning, Teaching, and Education

11 Simulated Practice
An Alternative Reality

Naomi Shiner

INTRODUCTION

This chapter explores the use of simulated practice within a general radiography context. General radiography is the initial modality learnt by students and is practiced in the majority of radiography departments. The diversity offered by each individual patient and condition requires a general radiographer to be able to use their knowledge and skills, tailoring it to every individual patient. In most cases, this is undertaken in a very short time frame.

Lower radiation doses, lack of contrast agents, drugs, and strong magnetic fields means general radiography can be considered the safer of the modalities available within Radiography. Simulated practice plays a critical role in improving safety within healthcare (Nestel et al. 2018). It can also be costly both financially and with time. So why would you consider simulation as a pedagogical tool in this context?

The traditional approach used in simulation: briefing, intervention, and debriefing will be discussed separately. This is not written as a 'how to guide', but each section will include tables to demonstrate diversity and tips that can be used to help introduce simulated practice. Some theories that underpin simulated practice will be explored to understand its relevance and advantages over traditional teaching approaches in a healthcare profession. An example case study will be used to demonstrate the potential for improving patient experience.

The term 'facilitator' is used as an umbrella term describing any individual teaching another individual using simulation. The term 'participant' may represent a preregistered radiography student or qualified radiographer participating in a simulation.

BRIEF OUTLINE OF SIMULATION IN HEALTHCARE

Several definitions for simulation or simulated practice are presented within the literature. A widely used definition is presented by Gaba (2004, p. 2) *'A technique to replace or amplify real experiences with guided experiences, often immersive in nature, that evoke or replicate aspects of the real world in interactive fashion'*. This reflects the diversity offered by this pedagogical approach.

Simulated practice has been used to improve performance and has not been limited to any industry. Historically, simulated practice can be seen in many forms: gaming in the form of chess to home tactical skills and medieval jousting to improve accuracy on horseback (Bradley 2006).

Taking a broader perspective, it can be argued that healthcare-simulated practice has evolved over several centuries. Modeling anatomical features has been seen in carvings since 24,000 BC (Markovic & Markovic -Zˇ Ivkovic 2010), and more recently in Egyptian times (Olry 2000). However, consideration should be given to distinguishing what is simulated practice and what is art or sculpture. This line is blurred as it is the appreciation of human form within the art world that has led to developments in simulated practice. Owen (2012) identifies mediums used for simulated procedures in medical education; historically ranging from intentional beheadings of criminals in order to recreate and understand injury with cadavers, to use of wax works and papier-mache for modeling of anatomy and disease.

Pioneers in simulated practice such as Laerdal and Safar (developer of the Resuci-Anne a part-task trainer, still readily used today across the world to support Cardio Pulmonary Resuscitation (CPR)) recognized value in providing experience and repeated practice to prepare individuals prior to encountering a real patient (Cooper & Taqueti 2008). Development of manikins and part-task trainers (permits the practice of a psychomotor skill in isolation, e.g., an artificial arm for venepuncture or X-ray phantom of a limb) allowed for repeated simulated practice. It is the ability to repeat a process without causing harm to real patients that embeds simulated practice as an ideal educational tool, improving patient care and experience.

THE ROLE OF EXPERIENTIAL LEARNING AND SIMULATED PRACTICE

General radiography requires practical application of a broad knowledge base, and the ability to adapt and innovate to every patients' individual circumstances. Thus, the education of radiography students has roughly a 50:50 split between academic and clinical placements in the United Kingdom (UK). Didactic lectures are still commonplace in the academic curricular and are an appropriate option to teach relevant theory to students prior to undertaking simulated practice. Simulated practice is considered to be a good medium to bridge the transition of theory to clinical practice (Hyde & Strudwick 2017). This is relevant to both pre- and postregistered radiographers as new developments are frequent within the profession. This section will concentrate on some educational theories used to support the use of simulated practice.

Traditional teaching approaches such as 'see one, do one, teach one' assumes every individual gains confidence and competence through the same level of exposure to practice. This assumption has inherent risks, adding undue pressure on individuals to practice without enough training and/or confidence, and, in turn, placing patients in a vulnerable position. Human factors are recognized as a cause of error or incidence within health care, with simulation seen as a tool to address these problems (Anderson et al. 2014).

Simulation is an experiential approach to learning. John Dewey introduced the theory of experiential learning (Miettinen 2000). Dewy believed that education experiences are an interaction between an individual and prior experiences. Kolb was heavily influenced by Dewy and developed the learning cycle, essentially a cycle of learning and relearning (Kolb 1984). This requires full engagement in an experience, observing, reflecting, creating new concepts, and making decisions to solve problems. When a radiography student or radiographer has little prior experience to draw upon some of these aspects may be challenging. Individuals responsible for education within a radiography setting can support students and staff by adopting Knowles's conceptual theory of 'Andragogy'. This theory is based on characteristics that separate adults from children learners (Knowles 1968). It is largely based on a learner centered approach with seven strategies supporting adult learning. Combined the theories presented by Dewey, Kolb, and Knowles reflect many of the principles identified in the six phases of simulation, outlined further in the chapter. Following are seven steps based on Knowles's theory of andragogy to facilitate adult learning:

1) Provide a safe and comfortable environment to encourage self-expression.
2) Provide opportunity for learners to identify their own learning needs.
3) Engage learners in the planning of their education.
4) Learners to identify their own learning objectives.
5) Learners to identify resources and devise their own strategies to meet their learning objectives.
6) Provide support.
7) Involve learners in self-evaluation and critical self-reflection.

To be a registered radiographer the Health Care and Professions Councils (HCPC) (UK) requires radiographers to complete a range of activities as part of their continued professional development (CPD) (HCPC 2017). Employers and radiographers are encouraged to draw upon these seven strategies to support engagement with this process. Adopting simulated practice engages the adult learner in an immersive experience pulling on both the cognitive and emotional forms of learning. Life experience shapes an individual's actions. Feelings, knowledge, assumptions, personality traits, and relationships inform these actions. As such, adult learners will each perform differently. Adults learn best when the activity is linked closely to their real world (Simons 1999).

General radiography requires individuals to combine technical and softer interpersonal skills. This combination can be difficult to learn or refine in the presence of a real patient, the situation can become uneasy and awkward that can damage an individual's self-esteem if there are poor outcomes. Confidence is key in general radiography to gain the trust of cooperation from patients (PPPP 2018). Simulation offers the facility to steer, stop, pause, and restart events unlike a real-life situation. This provides a safe learning environment for individuals to explore their skills and knowledge, testing themselves in new challenging situations preparing for the future.

Postregistration radiography department-based education often takes the form of short 'lunch time' lectures, mandatory training, or online learning. Time and skeletal staffing are often the limiting factors allowing CPD to take

place within the workplace. Management must therefore consider the investment versus impact required to facilitate simulated practice within the workplace. This relationship will be considered in a later section. Below are several
drivers for using simulated practice within general radiography education:

✓ Improve health and safety for staff and patient.
✓ Manual handling procedures, for example, practicing procedures such
  as recovering a fallen patient.
✓ Ethical considerations avoiding harm to patients by not training on
  them.
✓ Increased demand on services limiting training time in clinical setting.
✓ Quality of education.
✓ Duty of care to learners and clinicians.
✓ Rigorous assessment.
✓ Fast-changing developments in radiographic equipment.
✓ Developments in technology and specialist centers facilitating access
  to simulators and simulation.
✓ Interprofessional team building.
✓ Improved patient-centered care in time-limiting conditions.
✓ Developing service improvements.
✓ Test new protocols and procedures such as major incidents.
✓ Restructuring departments.
✓ Leadership training and development.

REALISM

The term 'realism' is often used when discussing simulations. A definition of
realism is '*the attitude or practice of accepting a situation as it is and being prepared to deal with it accordingly*' (Leixco.com 2019). This term also links well
with fidelity. Simulations are often referred to as having high or low fidelity.
Fidelity is linked to the likeness of a situation. The term 'high fidelity' maybe
used when efforts are made to replicate a real environment and situation.

Learning draws upon all five senses: touch, sight, sound, taste, and smell.
Within a radiography setting, each of these are used to inform judgement,
behavior, practice, and to support the patients during imaging procedures.
Integrating the use of each sense increases the fidelity, allowing individuals to
behave as they would in a real situation improving the transfer of skills and
knowledge from simulation to practice.

Facilitators can use several techniques to add to the realism or fidelity. Availability of setting may dictate the level of work required to add to the realism. In-
situ simulations held in an X-ray room or in resuscitation addresses many of the
senses by offering real background noise opposed to recorded noise in a Higher
Education Institute (HEI) setting. Patients may be simulated using higher end
simulators, such as SimMan 3G, which are programmed to have physiological
metrics and responses mimicking normal or abnormal disease. These mannequins
are limited where interpersonal relationships are required as they lack nonverbal

expression and physical movement. This can be introduced by using actors to role play as the patient. Here, the situation may 'feel' more natural allowing participants to interact more readily. In this scenario, pure use of a simulated patient means assessment of this individual would fail to offer the physiological metrics offered by a mannequin. Both offer elements of realism; therefore, it is important to consider the main learning outcomes intended for the simulation to guide the planning stage.

Dieckmann (2009) suggests simulations offer their own reality. Individuals are required to act or improvise within the simulation. For some suspending disbelief empowers them to work through the situation. Others may find this overwhelming and a distraction from the task in hand.

Facilitators must consider at some stage a participant may not know what 'a real situation' looks like. This is more likely the case with first-year radiography students prior to clinical placement. The simulation can lack context for these students; in some instances, the suspension of disbelief means an individual's experience becomes their reality. This can be counterproductive placing them in a more challenging and possibly confusing situation when presented with a patient in a real setting. Facilitators should use preparatory work and briefings to address this situation.

Engagement in all phases of a simulation is required to promote a meaningful experience. Meaningfulness is not reliant on high-fidelity equipment or high levels of realism. Meaningfulness stems from understanding. Discussions should include purpose, feelings, relationships within the team (interactions), performance (self and others) decision-making, and application to future events.

Tip: Engage participants from the beginning, ask them what would make it a meaningful experience. What are their personal objectives?

Tip: When resources that add to realism are not available, use an item to represent another enabling interaction to take place. Ensure everyone understands what is in the briefing to avoid confusion.

PHASES OF SIMULATION

A simulation consists of six phases: planning, briefing, intervention, debriefing, reflection, and evaluation (Nestel et al. 2018). When constructing a new simulation, it is important to consider the purpose of the simulation. This may be a training exercise on new equipment, test of competence, exploring new procedures, or identifying strengths and weaknesses. The purpose will direct you to constructing the simulation. Briefing, interventions, and debriefings will be explored separately below. This will be followed by an example scenario highlighting the potential impact on the patient experience.

RISK ASSESSMENT

Prior to any simulated practice a risk assessment should be undertaken. The depth of this assessment will depend on the type of simulation and individuals

TABLE 11.1

Source or potential person at risk	Action to minimize risk
Participant	Briefing – Physically and mentally fit to be involved in task. Identify type and level of tasks and request completion of self-assessment. – Manual handling: Proof of current competence. – Psychological effect: Provide a debrief and direct individual to counseling services.
External personnel	Discuss and agree the number of scenarios in each period. Identify a 'safe word', which is agreed prior to simulation with the simulated patient, whereby they can use the word to indicate that they wish to stop the simulation at any point. Ensure there is a facilitator always present during the simulation to observe the simulated patient. Ensure the 'patient' is briefed fully and orientated to surroundings prior to the simulation and receives a debrief with regards to the simulation and their welfare. At all times the privacy and dignity of the 'patient' should be maintained with a presimulation agreement around the necessary clothing required, the requirement of the script and a private space to get changed if required.
Equipment	– In-service date. – Checked over for signs of damage. – Participants are trained to operate equipment prior to use. – Suitable equipment is provided for tasks to be undertaken. – Use of radiation only on phantom equipment.

involved. Any activity requires the identification of a risk, assigning a severity, and taking action to minimize the risk where possible. Table 11.1 provides examples of some risks and steps considered to minimize negative impact.

BRIEFING

A briefing can be defined as *'information or an orientation session held prior to the start of a simulation-based learning experience in which instructions or preparatory information is given to the participants'* (Meakim et al. 2013, p. 6). This may include several activities that are tailored to the purpose of the simulation. Some simulations require participants to undertake preparatory work in advance of the simulation; Table 11.2 provides some examples. Specific activities are undertaken immediately prior to the simulation. Table 11.2 outlines these activities; each helps the learner/s to understand the purpose of the simulation-improving engagement.

As simulated practice is still a new experience for many individuals, it can promote anxiety related to individual performance (Tyerman et al. 2016). Lack of experience in simulation has been associated with increased stress

TABLE 11.2

Briefing guidance

Preparatory work	Briefing activities in advance of the simulation
Web-based learning	Introductions
Prereading	Identification of roles
List of skills to be mastered prior to attending	Orientation to environment
Creating cognitive aids, e.g., cue cards	Orientation to equipment
Attendance to mandatory training sessions to home psychomotor skills	Review of learning objectives of the simulation
Video	Overview of learner roles and expectations
Lectures	Opportunity for questions and answers
Self-assessment, e.g., quiz	Safe words

(Gouin et al. 2016). This can overwhelm, inhibiting them from performing and gaining any positive outcome from the experience. Repeated exposure and level of briefing is key to reducing this stress. Yet dependent on the reason for the use of the simulated exercise, that is, formative or summative, the depth of information provided to the learner may vary considerably. Either way, enough information should be provided to ensure the safety of all those involved.

> *Tip: Avoid trickery. A simulated exercise should have a clear learning objective and should not be designed to trick the individual. This only serves to reduce the trust between the facilitator and the participant reducing engagement and likelihood of a positive outcome.*
>
> *Tip: Provide sufficient time for all involved in any role play to become familiar with the details. This may require separation of individuals into different areas and opportunity for clarification prior to the intervention.*

INTERVENTION

The intervention or simulated activity can be widely diverse within a general radiography context. Its only limitation is the imagination. That being said, it should be appropriate to allow the learning outcome to be fullfilled.

As discussed earlier, high fidelity aims to be realistic immersing the participant in a reality that is more reflective of their working environment. Stokes-Parish (Stokes-Parish, Duvivier & Jolly 2017) discusses the role authenticity plays in achieving this within simulation, indicating the need for attention to detail. This ensures participants do not dismiss elements of a simulation that are less authentic and miss the relevance of the activity. An example of a higher fidelity simulation within radiography is presented later in the chapter.

At the opposite spectrum is low fidelity. This may be the use of part-task trainers rather than full interactive manikins or an individual role playing as the patient. It may be questioned why use low fidelity if high fidelity implies a better learning experience? Scaffolding is an educational concept used in simulation, where facilitators break down the skill or knowledge to be learnt into stages (Van Merriënboer & Sweller 2010). Facilitators support and encouraging participants to build on prior knowledge; as competence grows the need for scaffolding reduces. Radiography is a complex application of technical and interpersonal skills. Figure 11.1 demonstrates how scaffolding might be approached using a variety of simulated approaches identified in Table 11.3, culminating into a higher fidelity scenario reflective of a real working environment.

Figure 11.1 indicates complexity of a scenario may increase according to level of learning. Cognitive load theory aims to develop instructional design based around reducing load on a person's working memory. Too much load on working memory reduces the brains ability to transform knowledge into long-term memory. There are three types of cognitive load: intrinsic which is the complexity of the information needed to be processed; extraneous load is information that is not linked to the learning task but uses working memory and is therefore a distraction; lastly, germane load that is mental effort that supports the development of schema in the long-term memory. Schema are describing a pattern of thought or behavior that organizes categories of information and the relationship among them. Individuals use schemata to help understand rapidly changing environments by influencing their attention and absorption of new knowledge. This builds new frameworks to support future learning.

Scaffolding, rehearsing, and sequencing all support germane load and are found in simulated practice. A facilitator should aim to minimize intrinsic load by simplifying information working from low to high fidelity; minimizing extrinsic load by maximizing engagement of participants and using goal-free tasks, worked examples, and integrating different sources of information; and

TABLE 11.3
Approaches to simulated activities used in general radiography (Shiner 2018)

Simulated activity	Objective
Gaming	Enable understanding of professional roles and responsibilities in care settings.
Video/DVD recorded roleplay	Enhance awareness of the patient–practitioner interaction.
Role play	Improve interpersonal skills
Computer software	Explore technical skills.
Virtual reality	Proficiency in image quality assessment.
In-situ	Preparedness for trauma situations and interprofessional practice.
Integrated actors	Improve knowledge of different professional roles and interprofessional working.
Manikins, phantoms, or part-task trainers	Focus on collaboration, problem solving, and empathy.
Simulated suits	Improve attitudes toward older adults and certain conditions.

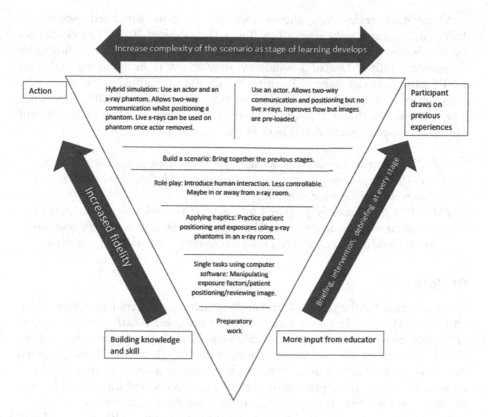

Increase complexity of the scenario as stage of learning develops

Action

Hybrid simulation: Use an actor and an x-ray phantom. Allows two-way communication whilst positioning a phantom. Live x-rays can be used on phantom once actor removed.

Use an actor. Allows two-way communication and positioning but no live x-rays. Improves flow but images are pre-loaded.

Participant draws on previous experiences

Build a scenario: Bring together the previous stages.

Role play: Introduce human interaction. Less controllable. Maybe in or away from x-ray room.

Applying haptics: Practice patient positioning and exposures using x-ray phantoms in an x-ray room.

Single tasks using computer software: Manipulating exposure factors/patient positioning/reviewing image.

Preparatory work

Increased fidelity

Briefing, intervention, debriefing at every stage

Building knowledge and skill

More input from educator

FIGURE 11.1 Relationship between scaffolding, fidelity, and implementation of example simulations.

lastly, maximize germane load by increasing variability, including contextual interference whereby adjacent tasks require different skill sets and encourage self-explanation (see debriefing). Simulation provides opportunity for facilitators to draw on cognitive load theory when planning their intervention to maximize learning opportunities.

The stage of learning should be a key consideration. Preregistration students have less expertise in radiography than those in a qualified role. Simulations planned throughout a curriculum and within a qualified setting should take into consideration the 'expertise reversal effect'. With scaffolding, facilitators may provide simpler instructional methods which work well with a novice learner. When experienced individuals have more knowledge to draw upon, these instructional methods become obsolete or have a reverse effect and act as an extraneous load distracting from the learning process. Depending on the learning objectives of a simulation, experienced learners will require less preparatory work, step-by-step guidance and should be working toward generating all the required steps themselves to complete a task.

A literature review has shown variation in how simulated practice is delivered in general radiography (see Table 11.3) (Shiner 2018). The challenges for delivering simulated practice will differ from a HEI to a radiography department setting. Use of low-fidelity simulations is likely to be higher in a HEI settings as the stage of learning requires more scaffolding. Radiography departments can use in-situ simulations offering a higher fidelity experience more readily, but this is challenged by room availability and time. Investment will be discussed in more detail later in the chapter.

> *Tip: Troubleshoot by running through the simulation in advance of delivery. Discuss potential variations, questions, and information participants require to fully engage with the task.*
> *TIP: As a facilitator, be prepared to 'pause' a scenario when risky actions are about to or are undertaken. Use these moments as a learning opportunity and hold a discussion. Then reset the scenario and allow it to continue.*

DEBRIEFING

What is meant by the term 'Debriefing'? A definition offered by Cheng et al. (2014 p. 658) '... *a discussion between two or more individuals in which aspects of a performance are explored and analysed with the aim of gaining insights that impact the quality of future clinical practice'*. Debriefing moves beyond one's own understanding of an experience. It embraces the help of a facilitator to work through a description of events, feelings, skills, and knowledge leading to outcomes that provide opportunity to improve for future events.

Some may consider debriefing to be like feedback. Although feedback tends to be a unilateral event, using predetermined standards to assess one's performance. A formal debrief model offers structure, embracing the full communication cycle, and focuses a person's natural processes to reflect on an experience. Without this formality, an individual may reflect in an unsystematic way or not at all. This can lead to more complex and dangerous situations if errors have occurred but not addressed. Table 11.4 provides a brief overview of some popular debrief models.

It is common for a debrief to be undertaken immediately after a simulation, when actions, feelings, and communications are still fresh in everyone's minds. The setting of the debrief can vary dependent on need for the space (an X-ray room) or length of time allocated to the debrief. Holding a debrief in the same setting as a simulation can mean props act as prompts for discussion. However, if the simulation has resulted in high levels of tension, it can be more effective to move away from that environment allowing feelings to diffuse.

In some instances, in-simulation debriefs occur with pauses and restarts throughout the simulation. A concern with simulation is the potential for negative learning to occur, whereby participants act incorrectly but it is not addressed by the facilitator. In-simulation debriefs reduce this occurrence as errors are discussed at the specific time point. Van Heukelom, Begaz and Treat (2010) found that in-simulation debriefs did not detract from the realism. However,

TABLE 11.4
Examples of debrief models

Type	Overview
Participant lead	Participants lead the debriefing with minimal guidance from the facilitator.
Diamond debrief (Jaye, Thomas & Reedy 2015)	Provides question prompts as scaffolding for the facilitator debriefing. – Description – Analysis – Application of learning
Promoting Excellence and Reflective Learning in Simulation (PEARLS) (Cheng et al. 2016)	Offers a structured framework for facilitators learning to facilitate debriefings in simulation. – Learner self-assessment – Facilitating focused discussion – Provide information in the form of directive feedback and/or teaching
'Thinking on your feet' (Krogh, Bearman & Nestel 2016)	Themes gathered from experts in the art of debriefing. – Values – Artistry – Techniques – Development
Debriefing Assessment for Simulation in Healthcare (DASH) (Brett-Fleegler et al. 2012)	This is a tool aimed to assess a facilitator's ability to debrief. As debriefs are integral part of learning, evaluating all aspects is essential. This includes the facilitators role. – Sets the stage for an engaging learning environment. – Maintains an engaging context for learning. – Structures debriefing in an organized way. – Provokes interesting and engaging discussions and fosters reflective practice. – Identifies performance gaps. – Helps close performance gaps.

participants had a preference for postsimulation debrief with minimal in-simulation debriefing, allowing them to work through the scenario more naturally in real time. This may be more favorable in a general radiography setting where patient facing time is brief. In this instance, it is imperative the facilitator notes key issues to be addressed to avoid negative learning.

On occasion, there is value in holding a delayed debrief sometime after the simulation. This permits the reflective process to take place and is particularly valuable if participants are anxious, making it difficult to work through feelings. Participants can take a defensive stance during an immediate debrief. This is known as the 'conflict triangle' and is discussed later.

Involving all parties within a debrief offers greater perspective and context to the discussions. This can be valuable in understanding and improving the

patient experience. Employing 'experts by experience' (individuals with previous experience being cared for or in a caring role) to role play as a patient or
to undertake an observational role adds to the debrief.

Stage of learning, previous simulation experience, and purpose of simulation (formative or summative) should inform the type of debriefing and the
timing of its delivery.

> *Tip: Be generous with allocating time for the debrief. This is the phase
> that can truly bottom out issues and be a creative space to make
> a difference for radiographers and patients.*
> *Tip: Allocate observational roles to help manage larger groups of people.
> These participants will learn through observation and are pivotal in pro-
> viding peer feedback during the debrief stage.*
> *Tip: Level of facilitation to debrief will differ from group to group. Avoid
> temptation to 'over instruct' when participants are less open with their
> discussion. To promote engagement a good briefing is essential.*
> *Tip: Consider breaking larger groups into smaller subgroups for a self-
> debrief to then re-join the larger group to express their thoughts.*

FEELINGS AND CONFIDENCE

Confidence in a learner stems from the hope that future experiences reflect previous ones. Simulation offers the opportunity to experience new situations therefore they are familiar and manageable. Within each individual is a three-step
loop known as the 'conflict triangle' (Kets de Vries 2007). The three points consist of conflict (anxiety), defence mechanisms, and feelings. Each impacts on the
behaviors and character traits of an individual. Performance issues can affect
confidence and a radiographer lacking confidence fails to reassure a patient.

Feelings are triggered unconsciously in the body. The five primary feelings
are: happiness, sadness, anger, guilt, and love. Feelings may be superficially considered in a debrief by a participant. Yet what should be considered is the level
of physiological energy that these feelings emit. The unknown or fear of an
experience may promote an imposter syndrome response. This can create performance anxiety, designed to cover up the initial undesired feelings. Considering this in the context of an error at work, radiographers have a tendency to
think it's their fault due to a lack of knowledge and ability. Radiographers may
react in two ways following a perceived failure: displacement or projection.

Displacement is the placing aside of any energy linked to feelings such as
fear, anger, or aggression. This is done as it's not appropriate to discharge this
energy in the work environment as this may be perceived as unprofessional.

'a wish to conceal'

Although professionalism may favor displacement, the radiographer may
discharge these displaced feelings at home where the resources are not available to unpick the situation with those that help us to learn from the error.

This may have a negative impact on the radiographer's mindset contributing to burnout.

Projection sees a hidden feeling such as inadequacy build to anxiety, the radiographers defence in this situation is to project this energy onto someone else.

'a wish to reveal'

This can be perceived as blaming others for the error. This scenario may unfold in high pressure environments such as resus or theater but enables the individual to complete the task, though may impact negatively on their professional reputation.

Defence mechanisms are universal and are not a character flaw. They can be likened to the immune system or psychological homeostasis. It is used to protect oneself from the consequences or feared consequences of unexpressed emotion. It can be a communication device to tell our peers that we are suffering and require help. Kets de Vries (2007) discusses the value of placing individuals in a safe environment where they can be challenged and supported without hidden agenda. This is reflective of a simulated environment. Understanding defensive actions as the facilitator debriefing allows you to support participants rather than seeing them as obstructive. How to help these individuals? Colleagues or students who demonstrate defensive behaviors require you to remove that 'point' from the conflict triangle, restructuring the challenge as an opportunity for individual growth. Consider the timing of feedback and listen. When an individual is emotional, avoid critical feedback. Try to 'strike while the iron is cold not hot'. All individuals involved in the simulation should be briefed to remain empathetic, supportive to each other by participating and exploring similar experiences. It is just as important in difficult situations that participants are helped to acknowledge; it may take some time to change behaviors or cultures. This may be particularly important within radiography departments. Debriefing has been discussed previously; however, alternatives are an option. Different levels of intervention to understand individual feelings are offered in the following points but are not discussed in further depth:

- Delayed debrief
- Mentorship
- Individual supervision
- Group supervision
- Therapy

It is important to support participants to understand what emotions are experienced with patients and how this manifest in the working environment. The five main elements of emotional intelligence are self-awareness, self-regulation, motivation, empathy, and social skills. Simulations offer an opportunity to explore each of these through intervention and debriefing stages. Improving emotional intelligence has positive impacts on radiographers ability to work as an effective team member, being compassionate, and empathetic toward patients (Arora et al. 2010).

EXAMPLE CASE STUDY OF A HIGH FIDELITY SIMULATION

The use of simulation and moulage in undergraduate diagnostic radiography education: A burns scenario (Shiner & Howard 2019).

Learning objectives:

- To explore the role of imaging a burns victim.
- To gain student perception of using Simulation Based Education (SBE).
- To enhance preparedness for imaging patients with complex care presentations.

Briefing:

Student radiographers were provided with limited briefing via a participant information sheet inviting them to participate in the imaging of a patient with complex care needs.

Intervention:

Ward setting based in the university campus. Role play was used with additional props to enhance the realism of the scenario. Key to this was the use of moulage (application of mock injuries) creating 2nd and 3rd degree burns. Medical equipment including nasogastric tube, oxygen, monitoring devices, and simulated hematuria present in a urinary catheter added to the scenario. Ambiance was supported by background noises. The patient was in obvious discomfort, restless, and noncommunicative.

Debriefing:

In-situ debriefing and peer to peer discussion was facilitated by an academic immediately following the intervention. The 'patient' played by an academic also shared their observations.

IMPACT

Students completed pre- and postsimulation questionnaires using a five-point Likert scale focused on understanding and preparedness. The intervention was captured using SMOTs cameras and a GoPro mounted on the patient's chest to capture the patient perspective. Focus groups were completed 3 months later to allow for a period of reflection, application of knowledge and skills, and for immediate emotions to settle.

This simulation aimed to enhance the preparedness of student radiographers to image complex care patients specifically those with burns. Students considered the simulation to have offered them exposure to this type of patient they previously hadn't encountered. Moulage was found to add to the realism and was linked to their memory recall 3 months later. The simulation uncovered 'distraction' issues when viewing the burns with difficulties relating to the individual behind the injury. Students noticed each other's adopted defensive body language and facial expressions, acknowledging the negative impact this has on building patient trust and detracting from patient satisfaction. This simulation engaged the students on an emotional, physical, and cognitive level, offering them a challenging experience but in a safe environment.

The impact of this experience made students think about the wider picture, beyond imaging, considering not only the physiological issues this patient may have but also psychological and their family. The debrief helped to unravel any feelings and issues, supporting students to become more self-aware and facilitate a more patient-centered care approach. Students shared how valuable this simulation was for their learning. This included regarding it as a resourceful experience to reflect on once they were with a real patient.

IMPROVING PATIENTS EXPERIENCES THROUGH SIMULATION

So far, the theory and development of simulation has been explored. This latter half addresses why it is worthwhile investing in using simulation as an educational tool in radiography.

A challenge general radiographers have is variation. Patients age, ability, condition, culture, language, mental capacity, and referral makes every patient different to the next. Each patient has their own personal expectations based on lack of or previous experience. A strong theme present in the Patient, Public, and Practitioner Partnership task and finish group (2018) to improve patient experience is communication. Shiner (2018) established three key areas that are addressed by the use of simulation in general radiography: use of computer software linked to technical skills, interprofessional education, and practitioner–patient relationship. Improvements in each of these areas offers potential to improve communication with patients; an ongoing challenge for radiographers given the relative short time frame spent with patients in general radiography.

The use of phantoms or computer software enables participants to become more proficient in technical skills, such as patient positioning, exposure manipulation, or post processing. Confidence built with equipment in a simulated environment translates to better practice, reduced doses to patients and staff, and more efficient use of time with the patient (Bott et al. 2009).

Interprofessional working can be challenging for radiographers as they are a transient member of the team. Developing interprofessional simulations are challenging on many levels: the radiography profession is small in comparison making group organization difficult; creating scenarios where all professions are utilized at a meaningful level to justify the time involved in planning and delivering the simulation; and bringing together multiple professions whether during academic teaching or in a hospital environment. Despite these, interprofessional simulations has a positive impact on radiographers understanding of the patient condition, experience, and contribution of other professional roles (Alinier et al. 2014; Brown, Howard & Morse 2016; Buckley et al. 2012).

Radiographers understanding of the patient perspective can be achieved through role play. Wearing 'simulated ageing suits', using moulage, or exposure to realistic patient pathways places participants closer to the patients' experience. This translates into improved assessment of the patients psychological and physical needs (Bleiker, Knapp & Frampton 2011), thereby improving empathy and allowing greater time for communication with the patient.

While it is evidenced that simulations improve teamwork and communication in the short term. Many studies of simulation education within radiography fail to look at the longer term impact. Miller et al. (2012) recognized that improvements found in a trauma setting are not sustained and therefore recommended a continuum of simulations to maintain these benefits. This supports the continual professional development of radiographers and is a sensible approach given the fast developments present within the radiography profession, both technically and with personnel.

INVESTMENT

This section discusses the varying types and level of investment needed to implement simulated practice. Impact has always been a difficult outcome to measure following simulated practice. With every new simulated exercise used, it is imperative it is evaluated to ensure the level of investment is appropriate to inform future practices. This is the case in both HIE and in a radiography department. Forms of evaluation used in radiography-simulated learning will be discussed.

TIME

Drivers for using simulated-based learning were mentioned in section 1.2. Increased demand on services places pressure on radiographers to perform accurate imaging in a limited time while still providing patient-centered care. One key issue with simulated practice is the investment in time that is required to implement the six phases. So why, if our radiographers are working under such time constraints, would we invest in using simulated practice?

As indicated in Tables 11.2, 11.3, and 11.4, simulation is a very broad subject area. It is not possible to state an exact quantity of time required to see impact within a radiography department. However, on reviewing a range of simulated practice in general radiography identified by Shiner (2018), a number of key recommendations for developing simulations are made to improve the success and therefore impact making the time invested more worthwhile.

- ✓ Integration into the curriculum or department with increasing levels of difficulty.
- ✓ Create a framework for evaluation
- ✓ Create clear learning outcomes.
- ✓ Always use a briefing and debriefing.
- ✓ Align simulation to realistic clinical practice.
- ✓ Use trained simulated patients.
- ✓ Involve clinical partners.
- ✓ Provide the facilitator with training.

Tip: To ease time pressures in each of the six phases,. seek support from dedicated services such as simulation technicians in HEIs or simulation centers often linked to teaching hospitals.

Tip: Keep clear documentation of each stage, resources, required, and evaluative procedures. This reduces the time required when repeating simulations in the future.

Tip: Work interprofessionally to build a repository of simulated practice.

FINANCIAL

Financial investment in simulated practice can be considerably high dependent on the setting and level of simulated practice wished to reach. Although time has been discussed separately, time costs money and human resources in terms of finance should also be considered. As simulated practice is gaining momentum, there is building evidence supporting the employment of specific simulation lead roles to facilitate its use.

As a rule of thumb, low fidelity equals lower cost and high fidelity higher costs. HEIs and the National Health service (NHS) are in different situations with regards to financial investment. The HEIs are likely to require a higher financial investment to provide students with realistic environments in which to practice such as X-ray facilities. The NHS are at an advantage as a radiography department is readily available; however, disruption to services is required to free these for training purposes. The following list identifies equipment and its comparative cost for simulated practice for general radiography:

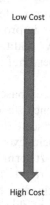

Low Cost

Moulage
Employing 'expert by experience' as actors (hourly rate)
Part-task trainers or phantoms
Age simulation GERT suits
Computer software packages, e.g., Shaderware (variable
　dependent on licencing arrangement)
Mannequins – SimMan 3G
Virtual reality software (variable dependent on licencing
　arrangement)
Immersive suites
Mobile X-ray unit
X-ray suite

High Cost

PERSONAL

Belief and engagement from all parties are required for the most successful outcomes to be achieved. A lack of experience with simulated practice can make some individuals sceptical of this pedagogical approach. Simulated practice often requires participants to role play; this repositioning of self can empower an individual to reach beyond their own limitations. This encourages freedom to test their knowledge and skills without fear of repercussion on a personal level. For others, the task of playing a role itself is uncomfortable and can act as a barrier to accessing their skills and knowledge. Overcoming these challenges lies in the preparation stages and debrief. Repeated exposure to simulated-based education results in more comfortable engagement and therefore greater impact on practice is achieved.

Tip: Create a program of simulated practice so participants have regular exposure and gain confidence in this pedagogy.

EVALUATION

The type of evaluation depends on the learning outcome. Three methods of data collection are available – quantitative, qualitative, and mixed methods. Determining which is best to support an evaluation should be led by the research question or aim of the simulation. For example,

1) Using imaging phantoms to simulate a new protocol or positioning with an aim to reduce radiation dose, would require collection of quantitative data.
2) Running a multidepartmental major incident simulation with an aim to review the pathway of mass casualties, a mixed methods may be appropriate. Gathering quantitative data such as length of time imaging and processing patients and qualitative data to explore staff opinion, providing opportunity to make the process more lean.
3) Interprofessional team work in accident and emergency with an aim to improve staff relations and understanding of each other's roles. A qualitative approach would be appropriate to gain descriptions, personal opinion, and personal account.

Evidenced-based research in simulated practice broadly but not solely favors confidence scales. These are often in the form of Likert scales using a 5- or 9-point scale distributed both pre- and postintervention, asking participants to rate their level of confidence against a skill set or knowledge. This evaluates the short term impact of the simulation. Frequently, a change will be seen in the shorter term demonstrating an improvement following a simulation. This is partly due to the freshness of the knowledge visited and the interactive nature of this educational approach. Given the time, financial, and personal investment in simulation, it is desirable that a longitudinal study is used to evaluate the longer-term impact. This type of evaluation also informs how often training should be revisited. This is a key principle of simulated practice, repeating and refreshing skills ready for infrequent events.

Evaluation is also key to revising simulations. Simulations can be complex, often improvements following the real event can be identified and solutions implemented. Addressing issues through evaluation ensures best practice is continually implemented. Simulated practice in the radiography profession is on the rise yet still has some development to match the established use in nursing and medicine. Disseminating the results and conclusions drawn from the evaluations helps the profession to learn from each other in a field that is still growing.

> *Tip: Evaluate the experience of those involved in the simulation too. This includes colleagues, simulated patients, and technicians. Different perspectives offer greater insight for improved practice.*

FUTURE CONSIDERATIONS

TRAINING

Evidence has shown the need for training for developing and facilitating simulated practice. (Thoirs, Giles & Barber 2011). Like any teaching approach, there is room for poor outcomes should the session not consider the audience, learning outcomes, and facilities used. Understanding and implementing the key concepts allows an individual or department to begin using this pedagogical approach. However, formal face-to-face simulation training provides opportunity for experience, question, and answer sessions supported by experienced facilitators. Courses are less readily available and can be costly both in finance and time. Other options are online learning platforms. These offer shorter, more accessible training courses, but lack the advantages offered by face-to-face training. A combination offering online modules with key workshops would be a good balanced training programme.

> *Tip: Access e-learning for health simulation resources.*
> *Tip: Contact local Simulation center to attend as an observer or facilitator to gain further experience and network.*

COLLABORATION

Investment within simulation can be both financially costly and time exhaustive. To overcome these and fully embrace the potential simulation offers, broadening approaches to accessing resources should be considered. Within the preregistration HEI setting, it is likely several resources are available that would support simulation. However, investment in new equipment and implementing simulated practice within the curriculum can vary from HEI to HEI (Shiner & Voyin 2019). It is here that collaboration adds valuable knowledge and greater potential for developing research opportunities. Cross HEI research can increase sample sizes in a profession that is characteristically smaller than nursing and medicine. Building a greater evidence base for this relatively new pedagogy within Radiography will support funding applications. With increased pressure on service, the need to grow the number of radiographers in training will require a curricular design that mirrors nursing curricular where simulation is valued as equal experience to clinical practice.

Simulation centers frequently have capacity and are under-utilized (Anderson et al. 2014). Radiology managers should embrace the expertise available at these centers to design and support simulations within radiography departments. Historically, radiology have limited involvement with these centers and are included almost as an adjunct to interprofessional simulations predominantly based in cross-sectional modalities. A common scenario is transferring adult-intensive care patients to a computed tomography scanner. As general radiography is exhaustively used throughout the hospital, as a profession we need to build on our inclusivity within this simulation community.

Lastly, who better to learn from but each other. Bringing pre- and postregistered radiographers together within a simulation would help to bridge the challenges in transition to clinical practice (Hyde & Strudwick 2017). Building relationships would improve understanding in current education techniques and developments within hospital settings. Simulation provides an opportunity to share experience and a voice, everyone has something to learn from each other despite their qualification status.

Tip: To improve collaborative opportunities join Simulation Interest Groups through the Society of Radiographer (UK).

Tip: Attend simulation-specific conferences to see latest innovative simulation resources, evidence base, and research design.

CONCLUSION

This chapter has provided a broad overview of the use of simulation as a pedagogical approach. It has highlighted the variation available to facilitators to build radiographers knowledge and skills within the profession. A six-phase approach is advised; however, the three key stages of briefing, intervention, and debriefing are discussed in more depth. Links between adult learning, realism, and cognitive loading have been highlighted indicating simulation is an appropriate choice for effective learning. The variation simulation offers, stage of learning, and setting means financial investment can also be variable.

Maximum success is dependent on engagement. Engagement moves beyond gaining experience within a scenario but also within all six-phases of the simulation process. Engagement means working with 'experts by experience' to learn how to make a positive impact; interprofessionally to build effective team working; and peers or colleagues to learn with, about, and from each other. Considered communication and the use of structured tools such as debrief models supports both the facilitator and the participants to learn about developmental needs and maintain professional skills and knowledge.

The profession is consistently seeing advances in radiography equipment and image analysis. This blended with the needs of the referrer and the wide variation of patients; radiographers require to continue with their education during preregistration and beyond qualifying. Simulation allows for exploration, challenging, problem-solving, development, and the building of competence and confidence, all within a safe learning environment. This translates into improved services and patient experience, which is at the heart of each radiographer's role.

REFERENCES

Alinier G. (2007) A typology of educationally focused medical simulation tools. *Medical Teacher* Volume 29 Number 8 Pages 243–250. DOI: 10.1080/01421590701551185.

Alinier G., Harwood C., Harwood P., Monatgue S., & Ruparelia K. (2014) Immersive clinical simulation in undergraduate health care interprofessional education: knowledge and perceptions. *Clinical Simulation in Nursing* Volume 10 Number 4 Pages 205–216. DOI: 10.1016/j.ecns.2013.12.006.

Anderson, Baxendale, Scott, Mossley, & Glover. (2014) The national simulation development project: summary report. Available at: http://aspih.org.uk/wp-content/uploads/2017/07/national-scoping-project-summary-report.pdf

Arora S., Ashrafian H., Davis R., Athanasiou T., Darzi A., & Sevdalis N. (2010) Emotional intelligence in medicine: a systematic review through the context of the ACGME competences. *Medical Education* Volume 44 Pages 749–764.

Bleiker J., Knapp K.M., & Frampton I. (2011) Teaching patient care to students: a blended learning approach in radiography education. *Radiography* Volume 17 Number 3 Pages 235–240. DOI: 10.1016/j.radi.2011.01.002.

Bott O.J., Wagner M., Duwenkamp C., Hellrung N., & Dresing K. (2009) Improving education on C-arm operation and radiation protection with a computer-based training and simulation system. *International Journal of Computer Assisted Radiology and Surgery* Volume 4 Number 4 Pages 399–407.

Bradley P. (2006) The history of simulation in medical education and possible future directions. *Medical Education* Volume 40 Number 3 Pages 354–362. DOI: 10.1111/j.1365-2929.2006.02394.x.

Brett-Fleegler M., Rudolph J., Eppich W., Monuteaux M., Fleegler E., Cheng A., & Simon R. (2012) Debriefing assessment for simulation in healthcare: development and psychometric properties. *Simulation in Healthcare* Volume 7 Number 5 Pages 288–294. PMID: 22902606.

Brown C.W., Howard M., & Morse J. (2016) The use of trauma interprofessional simulated education (TIPSE) to enhance role awareness in the emergency department setting awareness in the emergency department setting. *Journal of Interprofessional Care* Volume 30 Number 3 Pages 388–390 DOI: 10.3109/13561820.2015 1121216.

Buckley S., Hensman M., Thomas S., Dudley R., Nevin G., & Coleman J. (2012) Developing interprofessional simulation in the undergraduate setting: experience with five different professional groups. *Journal of Interprofessional Care* Volume 26 Pages 362–369.

Cheng A., Eppich W., Grant V., Sherbino J., Zendejas B., & Cook D.A. (2014) Debriefing for technology-enhanced simulation: a systematic review and meta-analysis. *Medical Education* Volume 48 Number 7 Pages 657–666.

Cheng A., Grant V., Robinson T., Catena H., Lachapelle K., & Kim J. (2016) The promoting excellence and reflective learning in simulation (PEARLS) approach to health care debriefing: a faculty development guide. *Clinical Simulation in Nursing* Volume 12 Number 10 Pages 419–428.

Cooper J., & Taqueti V. (2008) A brief history of the development of mannequin simulators for clinical education and training. *Postgraduate Medical Journal* Volume 84 Number 997 Pages 563–570.

Dieckmann P. (ed.) (2009) *Using simulation for education, training and research.* Lengerich: Pabst.

Gaba D. (2004) The future vision of simulation in health care. *Quality and Safety in Health Care* Volume 13 Number 1 Pages 2–10.

Gouin A., Damm C., Wood G., Cartier S., Borel M., Villette-Baron K., Boet S., Compére V., Dureuil B et al. (2016) Evolution of stress in anaesthesia registrars with repeated simulated courses: an observational study. *Anaesthesia Critical Care & Pain Medicine* Volume 36 Pages 21–26.

Health and Care Professional Council. (2017) Continuing professional development and your registration (online). Available at: www.hcpc-uk.org/globalassets/resources/guidance/continuing-professional-development-and-your-registration.pdf [Accessed on 19/10/2019].

Hyde E., & Strudwick R. (2017) How prepared are students for the workplace? *Imaging and Therapy Practice* Volume 9 Pages 5–11.

Jaye P., Thomas L., & Reedy G. (2015) 'The Diamond': a structure for simulation debrief. *The Clinical Teacher* Volume 12 Pages 171–175.

Kets de Vries M.F.R. (2007) Creating transformational executive education programs. *Academy of management Learning and Education* Volume 6 Number 3 Pages 375–387.

Knowles M.S. (1968) Andragogy, not pedagogy. *Adult Leadership* Volume 16 Number10 Pages 350–386.

Kolb D. (1984) *Experiential learning. Experience as the source of learning and development.* Englewood Cliffs, NJ: Prentice Hall.

Krogh K., Bearman M., & Nestel D. (2016) "Thinking on your feet" – a qualitative study of debriefing practice. *Advances in Simulation* Volume 1 Number 12. Available at: http://advancesinsimulation.biomedcentral.com/articles/10.1186/s41077-016-0011-4.

Lexico. (2019). Definition of realism in English. Available at: www.lexico.com/en/definition/realism

Markovic D., & Markovic -Ž Ivkovic B. (2010) Development of anatomical models-chronology. *Acta Medica Medianae* Volume 49 Number 2 Pages 56–62.

Meakim, C., Boese, T., Decker, S., Franklin, A. E., Gloe, D., Lioce, L., Sando, C. R., & Borum, J. C. (2013, June). Standards of Best Practice: Simulation Standard I: Terminology. Clinical Simulation in Nursing, Volume 9 Number 6 Pages S3-S11. http://dx.doi.org/10.1016/j.ecns.2013.04.001.

Miettinen R. (2000) The concept of experiential learning and John Dewey's theory of reflective thought and action. *International Journal of Lifelong Education* Volume 19 Number 1 Pages 54–72. DOI: 10.1080/026013700293458.

Miller D., Crandall C., Washington C., & McLaughlin S. (2012) Improving teamwork and communication in trauma care through in situ simulations. *Academic Emergency Medicine* Volume 19 Number 5 Pages 608–612.

Nestel D., Kelly M., Jolly B. & Watson M. (2018) *Healthcare simulation education.* West Sussex: Wiley Blackwell.

Olry R. (2000) Wax, wooden, ivory, cardboard, bronze, fabric, plaster, rubber and plastic anatomical models: praiseworthy precursors of plastinated specimens. *Journal of the International Society for Plastination* Volume 15 Pages 30–35.

Owen H. (2012) Early use of simulation in medical education. *Simulation in Healthcare* Volume 7 Number 2 Pages 102–116.

Patient Public and Practitioner Partnerships Task and Finish Group. (2018) Patient public and practitioner partnerships within imaging and radiotherapy: guiding principles (online) Available at: www.sor.org/learning/document-library/patient-public-and-practitioner-partnerships-within-imaging-and-radiotherapy-guiding-principles [Accessed on 29/09/2019].

Shiner N. (2018) Is there a role for simulation based education within conventional diagnostic radiography? A literature review. *Radiography* Volume 24 Number 3 Pages 262–271. DOI: 10.1016/j.radi.2018.01.006.

Shiner N. & Howard M.L. (2019) The use of simulation and moulage in undergraduate diagnostic radiography education: a burns scenario. *Radiography* Volume 25 Number 3 Pages 194–201. DOI: 10.1016/j.radi.2018.12.015.

Shiner N., & Voyin P. (2019) An overview of the types and applications of simulation-based education within diagnostic radiography and ultrasound at two higher education institutions. *Imaging and Oncology* Pages 6–13. Available at: www.sor.org/system/files/article/201905/io_2019_lr.pdf

Simons P.R.J. (1999) Transfer of learning: paradoxes for learners. *International Journal of Educational Research* Volume 31 Number 7 Pages 577–589.

Stokes-Parish J.B., Duvivier R., & Jolly B. (2017) Does appearance matter? Current issues and formulation of a research agenda for moulage in simulation. *Society for Simulation in Healthcare* Volume 12 Number 1 Pages 47–50.

Thoirs K., Giles E., & Barber W. (2011) The use and perceptions of simulation in medical radiation science education. *Radiographer* Volume 58 Number 3 Pages 5–11.

Tyerman J., Luctkar-Flude M., Graham L., Coffey S., & Olsen-Lynch E. (2016) Pre-simulation preparation and briefing practices for healthcare professionals and students: a systematic review protocol. *JBI Database of Systematic Reviews and Implementation Reports* Volume 14 Number 8 Pages 80–89. DOI: 10.11124/JBIS-RIR-2016-003055.

Van Heukelom J.N., Begaz T., & Treat R. (2010) Comparison of post simulation debriefing versus in-simulation debriefing in medical simulation. *Society for Simulation in Healthcare* Volume 5 Number 2 Pages 91–97.

Van Merriënboer J.J.G., & Sweller J. (2010) Cognitive load theory in health professional education: design principles and strategies. *Medical Education* Volume 44 Number 1 Pages 85–93.

12 Education Matters in Radiography

Iain MacDonald

KEY REQUIREMENTS IN RADIOGRAPHY EDUCATION AND THE CURRENT MODEL OF EDUCATION WITHIN THE UNITED KINGDOM

As all radiographers practising in the United Kingdom (UK) must be registered with the Health and Care Professions Council (HCPC), it is instructive to examine the requirements for registration with this body. An undergraduate course that a student completes must be approved by the HCPC to allow graduates to register as a radiographer, and the requirement is for a Bachelor's degree with honors. Within the requirements for a radiography course, Section 4.1 states that 'Learning outcomes must ensure that learners meet the standards of proficiency for the relevant part of the register' (Health and Care Professions Council, 2017). Essentially, therefore an undergraduate programme must satisfy the requirements of the standards of proficiency (Table 12.1).

Within these quite necessarily specific requirements there is leeway for those working within the teams delivering radiography education to use ideas around teaching and learning to constantly strive to improve the courses for students. With the introduction of tuition fees in the UK for radiography students in 2017, this has made the environment more competitive, with students becoming more discerning, and courses need to consider their own unique selling points (USP) (Nightingale, 2016). Sloane and Miller (2017) argue that the role of the radiographer is in 'a state of flux', with the exact role of radiographers varying greatly between employers. The curriculum must therefore be developed to match the current roles that radiographers undertake, given that there are very rapidly changing organizational, technological and social influences upon the role.

THE WAY STUDENTS LEARN IN UNDERGRADUATE EDUCATION

Before exploring detailed aspects of undergraduate education in radiography, it is useful firstly to give context, by exploring how learning takes place in more general terms, within the liberal adult education framework that is prevalent in the United Kingdom. This framework embraces individualism and autonomy with concepts such as self-determination or self-transformation as underlying philosophies (Jarvis *et al.*, 2003), and has produced a view of

TABLE 12.1

The Standards of Proficiency for Radiographers (adapted from Health and Care Professions Council, 2013)

1. be able to practise safely and effectively within their scope of practice
2. be able to practise within the legal and ethical boundaries of their profession
3. be able to maintain fitness to practise
4. be able to practise as an autonomous professional, exercising their own professional judgement
5. be aware of the impact of culture, equality, and diversity on practice
6. be able to practise in a non-discriminatory manner
7. understand the importance of and be able to maintain confidentiality
8. be able to communicate effectively
9. be able to work appropriately with others
10. be able to maintain records appropriately
11. be able to reflect on and review practice
12. be able to assure the quality of their practice
13. be able to understand the key concepts of the knowledge base relevant to their profession

adult learning as essentially self-planned or 'self-directed', rather than based on a pedagogic basis of instruction (ibid.). Malcolm Knowles, influential in the development of adult learning theory and practice, suggested that adults do not need teachers in the way children do, and are more capable of taking charge of their learning, popularizing the term 'andragogy' to define this (Knowles, 1975). Knowles's view was that adult learning, compared to that of children, is characterized by self-direction, with the idea of the teacher more as facilitator (Jarvis et al., 2003). The current 'orthodoxy' in higher education teaching is concerned with the ideas of constructivism (Johnston, 2010). Rather than simply being told facts, and memorization of these at a rather superficial level, the 'new discourse' values creating deeper meaning, through practical activities, reflection and judgement (ibid). This signals a move in radiography education away from traditional lectures to much more emphasis on group learning and other activities, often involving the virtual learning environment (VLE) such as Blackboard or Moodle, ideas that will be considered later in this chapter. Students have 'greater responsibility' for their learning in this environment, as it is a student-centered process and they often work collaboratively (Ellington, 2000). The move away from a teacher-centered approach to one that is student-centered can provide an additional focus for improving student retention on undergraduate courses (Thomas, 2008). This is particularly important in radiography, as there is a shortage of radiographers (College of Radiographers, 2017), and, once recruited onto a course as much as possible should be done to maximize retention of students.

Johnston (2010) defines the shift of emphasis in higher education from being 'taught' to 'experiencing learning' by active learning, which implies an encouragement of self-regulation of student learning processes combined with reflection, self-assessment, and effective peer or tutor feedback on perform-ance. Self-regulated learning (SRL) has emerged as an important construct in education, as it has become clear that one of the key issues in learning is the learners' ability to select, combine, and coordinate cognitive strategies in an effective way (Boekaerts, 1999). Self-regulation of learning is the degree to which students are responsible and active participants in the processes they use to learn (Zimmerman, 2008a). Self-regulated processes use a three-phase cyclical model (Zimmerman & Moylan, 2009). The model has three central components or phases, involving 'forethought', 'performance', and 'self-reflection' (ibid). Goal setting and planning are examples of forethought. So, for example, if a first year student radiographer is to give a presentation on radiography of the hand, a well-worked plan should be in place for this – what do they want to achieve? This sets the scene for the performance, where strategies designed to attain the goals are implemented – in the example, actu-ally delivering the presentation on hand radiography. Self-monitoring during the performance stage provides feedback that is then assessed during the self-reflection stage – in effect answering the question, how did I perform? In the self-reflection stage, attributions that are controlled and uncontrolled are evaluated, and ideas to affect forethought goals now complete the cycle of self-regulated learning via the feedback loop; therefore, self-reflective processes affect forethought. This cyclical model demonstrates that there is a cumulative effect of self-regulation in learning, increasing a learner's skill and self-efficacy. They become more interested in the task and feel more self-efficacious in how they meet the learning tasks and goals (Zimmerman, 2008b). Within radiography education, the explicit recognition of processes of self-regulated learning should certainly be considered and then communicated to students as a guide to how to learn effectively from activities. To guide lecturers in how to achieve this, Pintrich (1999) identifies three key characteristics that are important for students to self-regulate their learning. First is self-efficacy: students who believe they can learn and are confident in their skills are more likely to self-regulate learning. This per-sonal resource can be drawn on when faced with difficulties and time-consuming tasks. Second is task value beliefs: students who find their coursework interesting and useful are more likely to self-regulate their learning. There is recognition that this is due largely to the coursework prepared by the lecturing team, that it must suit the learning objective of the module being taught and capture the imagin-ation of students. Finally, goal orientations: if a learner's goal is self-improvement and learning, termed 'mastery goal orientation', they are more likely to engage in the activities of self-regulated learning to improve their learning and comprehen-sion of the topic area. Similarly, a goal of bettering others in attainment 'relative ability goals' can have a positive influence on self-regulation to encourage the learner to deep levels of engagement. As suggested by Pintrich (1999) changes in the approach to classroom instructional practice in radiography education by highlighting to students the importance of self-regulation will result in more deeply engaged and self-regulating students, and educators should certainly con-sider this approach.

Aside from self-regulation in learning, researchers have previously stated two positions on the way students learn in higher education: students may tend towards *deep* or *surface* learning approaches (Biggs, 2003; Entwistle, 2009). The terms were coined in a Swedish study by Marton and Säljo (1976) regarding the learning of undergraduate students and have been widely used since, particularly in higher education, forming the foundation of much research in the field (Webb, 1997). Deep approaches are associated with better academic outcomes (Entwistle, 2009). Those with surface learning tendency often memorize information and can easily be disconcerted by any questions in examinations that require a fuller understanding of the material. They can cut corners in their study and may want to achieve a minimal pass, as they can possess a 'meal ticket' view of university as a stepping-stone to a career; they may have other priorities aside from academic ones that are affecting their engagement, together with or because of too high workload and insufficient time. They may consider their factual recall adequate, misunderstanding the requirements of higher education, have a cynical view of education, or display high anxiety (Biggs, 2003). The alternative approach to study, and one that is encouraged, is the deep approach to learning, where students want to understand it for themselves, grasping more clearly the author/lecturer's meaning. They are interested in the subject and want to extend their depth of understanding, which is encouraged in higher education. They do not cut corners, and approach tasks appropriately and with a meaningful attitude. They are curious and may be determined to do well, having good background knowledge of the subject. In higher education, Baeten *et al.* (2010) identify that students who are satisfied with the quality of the course, including the appropriateness of workload/assessment, teaching and clear goals are more likely to display deep learning characteristics. Further student attributes that can indicate deep approaches are older students and students whose personality is characterized by openness to experience, extraversion, conscientiousness, agreeableness, and emotional stability. In addition, if students are intrinsically motivated, feel self-confident and self-efficacious and prefer teaching methods that support learning and understanding, a deep approach will be more frequently adopted (Baeten *et al.*, 2010). Clearly, there are a multitude of factors in students' personalities that would promote a deep or surface learning approach. The main difference between surface and deep learning approaches is the intention to simply reproduce the study material or to *understand* it (Biggs, 2003); and most teaching staff in radiography education will have met these two types of students, with their differing approaches to learning.

After considering the benefits of deep learning, the question arises: is there anything that can be done by lecturers, when designing and delivering courses in radiography, to engender an environment that fosters deep learning and discourages surface learning. Avoiding a heavy workload and inappropriate assessment for students are important as this can promote a surface learning approach (Lizzio *et al.*, 2002). Gow and Kember (1993) defined two different types of academic departments: those concerned with 'knowledge transmission' and those with 'facilitating learning'. Those with knowledge transmission discouraged deep learning approaches, while those who facilitated learning were less likely to induce surface approaches. Newble and Entwistle (1986) also discuss the departmental teaching approach in the context of medical education, a field of education that has some commonality with radiography education. Some departments hindered, rather than assisting students in developing the desired deep approach, with assessment highlighted as an area to review to encourage deep learning. Another aspect in radiography to consider is the strong disciplinary nature of the course, as it is focused largely on preparing students to enter the radiography profession. Warburton (2003) identifies that if the interests and background of students have this strong disciplinary focus, rather than a broader course of study, deep-learning habits can be inhibited. A conclusion from this discussion is that departmental teaching approaches *can* have bearing on the learning approach adopted by students, as well as the students own personal preference for learning though the strong disciplinary focus of radiography does mitigate against this somewhat.

The Use of Assessment

Assessment has been identified by Entwistle (2009), as a driving force of studying; affecting the effort given by students and it guides the learning of students. Entwistle argues that organized effort in study can apply equally to surface and deep approaches and needs to be seen together with the approach to learning. Good levels of organized effort combined with a surface approach may be providing a satisfactory level of performance – but only early in the course. As tasks and assessment changes, this approach is increasingly less effective when understanding becomes a more important criterion in assessment, only a deep approach combined with organized effort will be consistently rewarded (Entwistle, 2009). From a radiography student's point of view, therefore, surface learning using rote learning may have some merit for learning the names of structures within the body – for example, the bones of the skeleton in the early course stages. It may be less successful when analyzing and comparing the many different approaches to the diagnosis of disease in later stages of the course.

Students expect to be assessed in higher education, and assessment 'shapes the experience of students and influences their behavior more than the teaching they receive' (Bloxham & Boyd, 2007). The authors go on to say that tutors instinctively know the importance of assessment, and that anecdotal evidence points that to 'a large extent' assessment activity is *the* learning activity. Only in assessment tasks students do 'seriously engage' with learning material – be that lecture notes or their reading. Careful thought does need to go into the design of assessment, as it can interfere with the deep approach to learning and influence the approaches students adopt; it is students' *perceptions* about the assessment that affect the quality of learning taking place and students should see the purpose of assessment in the way intended by staff (Entwistle, 2009). The quality of the feedback received on their work is of 'paramount importance' (ibid) and should be used in formative assessment, defined as an activity that occurs *during* a module providing information to staff and students on their progress (Bloxham & Boyd, 2007). Black and Wiliam (2009) also discuss ideas around formative assessment and that it should develop both self-regulated learning and involve discourse within the classroom. It should also help produce self-regulated learners (Nicol & Macfarlane-Dick, 2006; Clark, 2012). Learners themselves also generate internal feedback while carrying out their academic work. In the example considered earlier, with a first year student giving an oral presentation on radiography of the hand, they often have feedback from the lecturer involved in the session; it is also likely that their peers provide informal feedback and they generate their own internal feedback on their performance developing their ability to self-regulate their learning. A further benefit of feedback is allowing lecturers to see where difficulties are experienced and where teaching may be focused

Having established the importance and methods of obtaining good feedback in formative assessment, concepts of assessment that are useful for lifelong learning are explored in Boud and Soler (2016). They review Boud's previous work (Boud, 2000) on 'sustainable assessment'. This is a design of assessment that focusses on the contribution assessment has to learning that extends *beyond* the timeframe of a particular course of study. It is envisaged that the skills and knowledge gained during assessment, including skills in self-regulation and self-assessment, should be of great benefit in student's future career as radiographers. Sustainable assessment is built on the notions of the nature of assessment *of* learning moving to assessment *as* learning. It provides students with an approach to their future work practices, using a lifelong learning model, which includes them undertaking assessment of their own practice. Much conventional assessment in higher education does not provide the tools do this, Boud and Soler (2016) argue. Their view on the value of teacher feedback is quite powerful – learning is *not* sustainable, they argue, if it needs continuous information from teachers on students' work. Students should develop the ability to self-assess more effectively, rather than relying on others. Their conclusions raise some important questions that need to be addressed if sustainable assessment is to be designed by radiography lecturers for their students (Table 12.2).

THE VALUE OF GROUP ACTIVITIES

As an example of sustainable assessment, collaboration when carrying out assessment is a valuable, and often, preferable way of working on any project, rather than students being individually assessed. Paavola *et al.* (2012) define that increasing levels of collaboration, as well as creative processes and new technologies are emerging trends in human learning and cognition. They go on to explain that modern society is complex, and people must combine expertise to solve challenging problems, not solvable solely by individuals. Certainly, in the radiography context, team working is used widely in practice, and can often be multi-disciplinary in nature. A core benefit of group learning is the additional social capital, a sociological term that acts as shorthand for

TABLE 12.2

Key Points to Address When Designing Assessment to Be Sustainable (adapted from Boud & Soler, 2016)

What features of the assignment and accompanying activity prompt consideration beyond the immediate task?

In what ways does engagement in the activity foster self-regulation?

How does the activity help learners meet challenges they will find in practice settings?

How is engagement in the current activity likely to improve the capacity of students to make effective judgements about their work in subsequent ones?

Are the educational benefits of the task likely to persist once the particular knowledge deployed in it can no longer be recalled?

Does the activity enable students to appreciate, articulate, and apply standards and criteria for good work in this area?

Does the activity enable students to demonstrate those course-level learning outcomes that relate to preparation for learning post-graduation?

the positive aspects of sociability accruing from such communities formed within the learning environment (Portes, 1998). The benefits include a feeling of belonging, ability to influence a group, and a shared emotional connection (McMillan and Chavis, 1986). More recent work by Laal and Ghodsi (2012) identified that collaborative learning brought caring, supportive relationships amongst students; higher achievement; increased psychological health, self-esteem, and social competence. Aside from the social benefits, many authors have acknowledged the benefits of learning together in order to learn from each other (e.g., Piaget, 1971; Vygotsky, 1978). Their ideas of cooperative learning between peers have been influential in education generally, including higher education. Slavin (1988) states that while there are many excellent benefits to cooperative learning, there must be a goal for the group, and individual accountability: there should be no ability to 'hide' behind the group project in an educational context. Race (2014) provides a useful summary of the benefits of collaborative learning (Table 12.3).

According to Felder and Brent (1996), there is often some negativity from students to group work when it is introduced, as they have expectations to learn directly from the teacher from early stages in their education. There may be some who complain about others not pulling their weight and having to explain ideas to other team members who may have difficulty in understanding. They go on to say that it can be awkward for lecturers to adopt group learning if they are new to it. However, the steep learning curve is worth persisting along, as the benefits of student-centered learning outweigh the difficulties. In considering the depth of learning, Felder and Brent suggest that learning as a group can develop students who think more deeply, have better attitudes towards their subject and towards each other.

TABLE 12.3

The Benefits to Students of Learning Collaboratively (adapted from Race, 2014)

Develop confidence in speaking, presenting, arguing, and discussing.

Oral communication skills are useful for job interviews and oral exams.

Can learn from each other, adding to what is learnt in class, and online resources.

Develop interpersonal skills, how to collaboratively work with others (interpersonal skills), and particularly difficult characters.

Reflect together on how the learning is going, and how they compare to their peers.

Learners deepen their learning by verbalizing, explaining difficult ideas and concepts to each other.

To achieve good group working, Bloxham and Boyd (2007) discuss the practicalities of implementing group work in higher education, and these ideas are useful for informing the methods used in the study for the collaborative assessment design. There should be high levels of collaboration and negotiations between members, or the benefits of group working may be lost, which may happen if a task can be completed by 'merely sharing the work out at the beginning and then combining each person's contribution at the end'. Group size is deemed to be important, with Bloxham and Boyd (2007) suggesting three to five as being an ideal size. Beyond this, it is more difficult to involve all members in discussion and arranging meetings and work is very difficult. Group mix is important, with Falchikov (2001) suggesting that mixing students of all abilities with the high achieving students increases the lower achieving students' eventual achievement. Bloxham and Boyd (2007, p. 57) state that in their view, tutors should allocate students to groups reflecting the 'reality' of most work teams, encouraging students to work with a wide range of people. They would therefore also encourage 'mixed ability' groups. More generally, when considering the role of the tutor, there are different concepts of the role in collaborative learning in colleges (Barkley *et al.*, 2014). Should they play a minimal role, allowing students to develop their own ways of learning, or give more direction: structuring learning tasks, monitoring progress, and providing interventions if students go off track? These are interesting questions, and, from the authors' experience, in general, students prefer more direction, certainly in the first year of the radiography programme.

Technology in learning

A good example of the use of technology in group learning is the wiki. The term wiki is taken from the Hawaiian word *wiki*, meaning 'to hurry' (Wheeler *et al.*, 2008). A wiki has been described by Parker and Chao (2006) as a 'web communication and collaboration tool that can be used to engage students with others in a collaborative environment'. It allows users to generate web content collaboratively and is potentially open to the public (such as

Wikipedia), though in a teaching setting it is normally not seen by the public and is private within a university virtual learning environment such as Blackboard or Moodle. An example of two such web pages from a group wiki is given in Figure 12.1, based on a second year collaborative project based around the diagnosis of common pathologies.

This type of technology-enhanced learning environment (TELE) has the potential to support self-regulated learning (Bartolomé & Steffens, 2006). Kitsantas and Dabbagh (2011) state that Web 2.0 social software tools, which include wikis and social networking sites, can be used by faculty to facilitate self-regulation by students such as goal setting, self-evaluation, and help seeking. Wikis could be especially useful as students are able to plan, reflect, and perform – ideas closely related to self-directed notions like thinking ahead about task goals and reflecting on their progress (ibid.). In the author's experience, students who do not like giving oral presentations favor wikis, with the

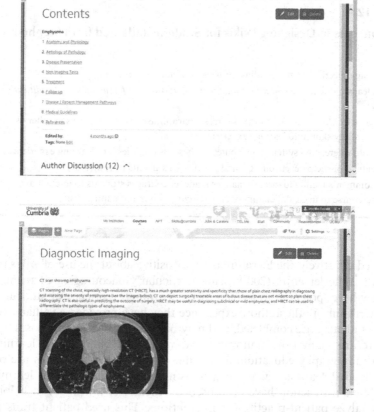

FIGURE 12.1 Example of second year undergraduate radiography wiki relating to diagnosis of disease (Authors own collection).

additional benefit that they are often able to contain more information than
Microsoft (MS) PowerPoint presentations. They allow users to build resources
collaboratively, giving a move towards a fully social constructivist form of
learning and have been used in diverse areas of higher education. In medical
education their use by undergraduates has been seen as creating links between
many different aspects of medicine, helping them understand the links between
them (Alireza *et al.*, 2009). It has also been used in the same field, outside
pure medical science, to create information and links between issues around
'professionalism' in medicine by allowing undergraduates to work together to
improve their knowledge, and pooling resources (Varga-Atkins *et al.*, 2010).
Wikis have been used in nursing education (e.g., Billings, 2009; Ciesielka,
2008), where they have found value in fostering collaboration, and group
work, though need to be well matched to the course of study (Zitzelsberger
et al., 2015). In radiography, they have been used in an interprofessional context
with other health care students (Stephens *et al.*, 2013). The authors proposed
useful ground rules for the effective use of wikis (Table 12.4):

TABLE 12.4

Key Concepts in Designing Wikis for Students (adapted from Stephens et al., 2013)

Use learning objectives as sub-heading: *provides structure to the wiki*

For each learning objective you have a maximum of 750 words: *forces students to edit rather than add words*

Do not paste blocks of text from other sources – paraphrase and reference as in other assignments: contributes study skills into learning experience

Use Harvard referencing system: contributes study skills qualities into learning experiences

All students to participate: encourages and permits all students to contribute

Use the comments facility to discuss changes made: encourages students to interact and engage in social intercourse to make difficult interactions – corrections and editing easier.

If used effectively, the literature is very positive about the use of wikis in learning, with Wheeler *et al.* (2008) being particularly encouraging of them, stating that they have 'potential to transform the learning experience of students worldwide'. Certainly, in the authors experience they have provided a valuable resource for the students once completed, and may be used for revision purposes.

There are many other innovative examples of technology in learning utilized in radiography education, and some of the diverse examples are outlined here. Blended learning environments using a mixture of online learning and face-to-face instruction have been used by Bleiker *et al.* (2011) in helping students analyze patient-practitioner interactions. This used patient transcripts of interactions, and then professional actors acted out the patient scenarios, thus preserving patient confidentiality. Accessing patient images for discussion and possible inclusion in case studies can be problematic for student radiographers, due to data protection measures. Cosson and Willis (2012) have managed to acquire over 50,000 individual patients' cases for their PACS system at their institution. This would seem to be to be a very valuable source of information for teaching, learning, and assessment. Macdonald-Hill and Warren-Forward (2015) in an Australian study found that by using techniques such as capturing lectures in video and allowing these to be seen online, together with online tutorials, allowed students to learn at their own pace. They were able to rewind lectures, which is not possible in traditional face-to-face lectures, with 94% of the students finding this valuable. These are just some examples of the use of technology to improve the learning experience for student radiographers.

CONCLUSION

This chapter has provided an overview of radiography education, and how it fits into the contemporary world of higher education. One method of analyzing how students learn, student approach to learning theory was used to highlight the different approaches students may have as undergraduates in radiography education. The requirement to design assessment to encourage the deep learning techniques that can improve academic outcomes for individual students and, ultimately, the profession as a whole, was considered. A method that can achieve deeper learning, as well as providing sustainable assessment, collaborative learning, was outlined, together with the increasing inclusion of technology-enhanced learning in the radiography curriculum. By concentrating on how the course is delivered in these ways, the needs of radiography students can be met more effectively in the rapidly changing world of clinical practice and education.

REFERENCES

Alireza, J., Margaret, M., Martin, G. and Lara, V., 2009. Wiki use and challenges in undergraduate medical education. *Medical Education*, **43**(11), p. 1117.

Baeten, M., Kyndt, E., Struyven, K. and Dochy, F., 2010. Using student-centred learning environments to stimulate deep approaches to learning: Factors encouraging or discouraging their effectiveness. *Educational Research Review*, **5**(3), pp. 243–260.

Barkley, E.F., Cross, P.K. and Cross, K.P., 2014. *Collaborative learning techniques: A handbook for college faculty*. New York: John Wiley & Sons, Incorporated.

Bartolomé, A. and Steffens, K., 2006. Self-regulated learning in technology enhanced learning environments. In: *VIII Congreso Iberoamericano de Informática Educativa*. San Jose, Costa Rica (13-05-2006).

Biggs, J.B., 2003. *Teaching for quality learning at university: What the student does*. 2nd edn. Buckingham: SRHE & Open University Press.

Billings, D.M., 2009. Wikis and blogs: Consider the possibilities for continuing nursing education. *The Journal of Continuing Education in Nursing*, **40**(12), pp. 534–535.

Black, P. and Wiliam, D., 2009. Developing the theory of formative assessment. *Educational Assessment, Evaluation and Accountability (Formerly: Journal of Personnel Evaluation in Education)*, **21**(1), pp. 5–31.

Bleiker, J., Knapp, K.M. and Frampton, I., 2011. Teaching patient care to students: A blended learning approach in radiography education. *Radiography*, **17**(3), pp. 235–240.

Bloxham, S. and Boyd, P., 2007. *Developing effective assessment in higher education. A practical guide.* Milton Keynes: Open University Press.

Boekaerts, M., 1999. Self-regulated learning: Where we are today. *International Journal of Educational Research*, **31**(6), pp. 445–457.

Boud, D., 2000. Sustainable assessment: Rethinking assessment for the learning society. *Studies in Continuing Education*, **22**(2), pp. 151–167.

Boud, D. and Soler, R., 2016. Sustainable assessment revisited. *Assessment & Evaluation in Higher Education*, **41**(3), pp. 400–413.

Ciesielka, D., 2008. Using a wiki to meet graduate nursing education competencies in collaboration and community health. *Journal of Nursing Education*, **47**(10), pp. 473–476.

Clark, I., 2012. Formative assessment: Assessment is for self-regulated learning. Educational Psychology Review, 24(2), pp. 205–249.

College of Radiographers, (2017). Diagnostic radiography UK workforce report 2017. Available at: www.sor.org/sites/default/files/document-versions/scor_census_of_uk_diagnostic_radiographic_workforce_2017_report_-_final_version.pdf (Accessed: 1st October 2019).

Cosson, P. and Willis, N., 2012. Digital Teaching Library (DTL) development for radiography education. *Radiography*, **18**(2), pp. 112–116.

McMillan, D.W. and Chavis, D.M., 1986. Sense of community: A definition and theory. *Journal of Community Psychology*, **14**(1), pp. 6–23.

Ellington, H., 2000. How to become an excellent tertiary-level teacher. Seven golden rules for university and college lecturers. *Journal of Further and Higher Education*, **24**(3), pp. 311–321.

Entwistle, N.J., 2009. *Teaching for understanding at university: Deep approaches and distinctive ways of thinking.* Basingstoke: Palgrave Macmillan.

Falchikov, N., 2001. *Learning together: Peer tutoring in higher education.* London and New York: Routledge Farmer.

Felder, R.M. and Brent, R., 1996. Navigating the bumpy road to student-centered instruction. *College Teaching*, **44**(2), pp. 43–47.

Gow, L. and Kember, D., 1993. Conceptions of teaching and their relationship to student learning. *British Journal of Educational Psychology*, **63**(1), pp. 20–23.

Health and Care Professions Council, (2013). *Standards of proficiency – Radiographers.* London. Available at: www.hcpc-uk.org/resources/standards/standards-of-proficiency-radiographers/ (Accessed: 1st October 2019).

Health and Care Professions Council, (2017). *Standards of education and training guidance.* Available at: www.hcpc-uk.org/globalassets/resources/guidance/standards-of-education-and-training-guidance.pdf (Accessed: 1st October 2019).

Jarvis, P., Holford, J. and Griffin, C., 2003. *The theory & practice of learning.* 2nd edn. London: Kogan Page.

Johnston, B., 2010. *The first year at university: Teaching students in transition.* Maidenhead: McGraw-Hill Open University Press.

Kitsantas, A., and Dabbagh, N., 2011. The role of web 2.0 technologies in self-regulated learning. *New Directions for Teaching and Learning*, **2011**(126), pp. 99–106.

Knowles, M.S., 1975. *Self-directed learning: A guide for learners and teachers.* New York: Association Press.

Laal, M. and Ghodsi, S.M., 2012. Benefits of collaborative learning. *Procedia – Social and Behavioral Sciences; World Conference on Learning, Teaching & Administration – 2011*, **31**, pp. 486–490.

Lizzio, A., Wilson, K. and Simons, R., 2002. University students' perceptions of the learning environment and academic outcomes: Implications for theory and practice. *Studies in Higher Education*, **27**(1), pp. 27–52.

Macdonald-Hill, J.L. and Warren-Forward, H.M., 2015. Feasibility study into the use of online instrumentation courses for medical radiation scientists. *Radiography*, **21**(3), pp. 282–287.

Marton, F. and Säljo, R., 1976. On qualitative differences in learning: I – Outcome and process. *British Journal of Educational Psychology*, **46**(1), pp. 4–11.

Newble, D.I. and Entwistle, N.J., 1986. Learning styles and approaches: Implications for medical education. *Medical Education*, **20**(3), pp. 162–175.

Nicol, D.J. and Macfarlane-Dick, D., 2006. Formative assessment and self-regulated learning: a model and seven principles of good feedback practice. *Studies in Higher Education*, **31**(2), pp. 199–218.

Nightingale, J., 2016. Radiography education funding – Crisis or opportunity? *Radiography*, **22**(2), pp. 105–106.

Paavola, S., Engeström, R. and Hakkarainen, K., 2012. The trialogical approach as a new form of mediation. In: A. Moen, A.I. Mørch and S. Paavola, eds, *Collaborative knowledge creation: Practices, tools, concepts*. Rotterdam: Sense Publishers, pp. 1–14.

PARKER, K. and CHAO, J., 2007. Wiki as a Teaching Tool. *Interdisciplinary Journal of Knowledge and Learning Objects*, **3**, pp. 57–72.

Piaget, J., 1971. *Science of education and the psychology of the child*. Trans. D. Coltman. London: Longmans.

Pintrich, P.R., 1999. The role of motivation in promoting and sustaining self-regulated learning. *International Journal of Educational Research*, **31**(6), pp. 459–470.

Portes, A., 1998. Social capital: Its origins and applications in modern sociology. *Annual Review of Sociology*, **24**(1), p. 1.

Race, P., 2014. *Making learning happen: A guide for post-compulsory education*. 3rd edn. Los Angeles, CA: Sage.

Slavin, R.E., 1988. Cooperative learning and student achievement. *Educational Leadership*, **46**(2), p. 31.

Sloane, C. and Miller, P.K., 2017. Informing radiography curriculum development: The views of UK radiology service managers concerning the 'fitness for purpose' of recent diagnostic radiography graduates. *Radiography*, **23**, pp. S16–S22.

Stephens, M., Robinson, L. and Mcgrath, D., 2013. Extending inter-professional learning through the use of a multi-disciplinary wiki. *Nurse Education in Practice*, **13**(6), pp. 492–498.

Thomas, L., 2008. Learning and teaching strategies to promote student retention and success. In: Crosling, G., Thomas, L., & Heagney, M. (Eds.), *Improving student retention in higher education: the role of teaching and learning* (pp. 69–81). Abingdon, UK: Routledge.

Varga-Atkins, T., Dangerfield, P. and Brigden, D., 2010. Developing professionalism through the use of wikis: A study with first-year undergraduate medical students. *Medical Teacher*, **32**(10), pp. 824–829.

Vygotsky, L.S., 1978. *Mind in society: The development of higher psychological processes*. trans. V.J.-S. M. Cole, S. Scribner and E. Souberman. Cambridge, MA: Harvard University Press.

Warburton, K., 2003. Deep learning and education for sustainability. *International Journal of Sustainability in Higher Education*, **4**(1), pp. 44–56.

Webb, G., 1997. Deconstructing deep and surface: Towards a critique of phenomenography. *Higher Education*, **33**(2), pp. 195–212.

Wheeler, S., Yeomans, P. and Wheeler, D., 2008. The good, the bad and the wiki: Evaluating student-generated content for collaborative learning. *British Journal of Educational Technology*, **39**(6), pp. 987–995.

Zimmerman, B., 2008b. Goal setting: A key proactive source of academic self-regulation. In: D. Schunk, & B. Zimmerman, eds, *Motivation and self-regulated learning theory, research, and applications*. New York: Lawrence Erlbaum Associates, pp. 267–296.

Zimmerman, B.J., 2008a. Investigating self-regulation and motivation: Historical background, methodological developments, and future prospects. *American Educational Research Journal*, **45**(1), pp. 166–183.

Zimmerman B. J., Moylan A. R., 2009. Self-regulation: where metacognition and motivation intersect. In: Hacker D. J., Dunlosky J., Graesser A. C. (Eds.), *Handbook of Metacognition in Education* (pp. 299–315). New York, NY: Routledge.

Zitzelsberger, H., Campbell, K.A., Service, D. and Sanchez, O., 2015. Using wikis to stimulate collaborative learning in two online health sciences courses. *Journal of Nursing Education*, **54**(6), pp. 352–355.

13 Educational Perspectives in Radiography

Julie A. Hendry

INTRODUCTION

As registered radiographers, part of our role includes supporting students to become future professionals. In fact the Health and Care Professions Council Standards of Proficiency state radiographers must *'understand the importance of participation in training, supervision and mentoring'* (2016, p. 8). Despite this, radiographers may feel ill prepared regarding their own skills and knowledge of pedagogic practice to enable the *'training, supervision and mentoring'* required by the profession. This chapter aims to provide a brief overview of the subject of pedagogy in radiography, which may be complicated as fee-paying students are now considered consumers of Higher Education (HE). Clinical radiographers have an essential role in learning and teaching alongside academic colleagues. This short introduction to the complexities of learning and teaching will hopefully be a catalyst for practitioners to further delve into the multifaceted world of student education. A variety of texts for further reading can be found at the end of the chapter.

BACKGROUND – THE LANDSCAPE OF 2020

Current students in the United Kingdom, who are becoming the radiographers of the future, face an exciting yet also very different workplace than that of the previous millennium. Introduction of fees and loss of bursaries, together with an NHS funding crisis and diminished workforce, all influence the education underpinning the radiography professions. We shall now briefly consider this changing landscape and its possible impact upon learning and teaching.

HIGHER EDUCATION

In August 2017, fees previously paid by Health Education England became the responsibility of prospective radiography students. The NHS bursary ended and students thereafter required loans to meet previously funded costs. As the Closing the Gap Report (King's Fund, 2019) explains, to minimize the difference between staff demand and numbers of graduates joining the professions, much significant investment, not a quick fix, is essential. With vacancy lists high, staff are struggling to meet the expectations of quality patient care, which also undoubtedly impacts professionals' ability to support and educate students. 'Burn out' is rising; 'there is no downtime' (Wilkinson, 2015, p. 842) reducing the opportunities for staff to reflect, teach, or ensure the delivery of high quality compassionate care to all. It might be argued that in over ten years since the failings in care reported by the Francis Inquiry (2013), little has been learnt.

The recent Augar Report (2019) into 'tertiary' or post-18 education in the UK suggests a reduction in fees for students, which could leave universities delivering high-cost medical education, with even greater financial problems. Moreover, this could negatively impact the learning and support afforded to radiography students as funding diminishes. Actions from Augar remain undecided but these limited example recommendations give context to the arguably problematic landscape of healthcare education.

STUDENTS OF 2020: GENERATION Z

Recent high turnover of newly qualified nurses and midwives within the Midlands prompted the Every Student Counts review. Study outcomes were published in the Health Education England (HEE) Mind the Gap Report (2015). The key message from the project was that between the generations of individual students and qualified staff involved, there are '... *differences in values, expectations, perceptions and motivations* ...' (HEE, 2015, p. 3). Although the action research study, which utilized focus groups and workshops, concerned nurses and midwives it transfers into the radiography setting. An awareness and comprehension of the differing motivational needs across generations would better enable employers and education providers to meet student (and practitioner) needs thus improving recruitment and retention. With this greater insight, student learning could arguably be improved, being more targeted to individual needs, preferences, and motivation.

Generation Z students, individuals born between 1995 and 2010, have begun to enter higher education. These '*Digital Natives*' (HEE, 2015, p. 13) are highly reliant upon mobile devices; seek instant communication and gratification, readily source information from 'Google' yet struggle to critique or apply this knowledge (HEE, 2015). When considering our approach to learning and teaching of these 'Digital Natives', it could be most effective when seen through a generational lens. Specific approaches to learning best suited to Generation Z will be considered after an explanation of what it means to learn and to teach – pedagogy.

THE CRAFT OF PEDAGOGY

Pedagogy is a frequently used term, but less often explored within particular healthcare education settings. The Oxford Online Dictionary defines pedagogy as '*The method and practice of teaching, especially as an academic subject or theoretical concept*' (Lexico.com, 2019, pedagogy entry). But how does this translate to the clinical radiography setting, and what does the all-encompassing word 'teaching' mean? The relationship between learning and teaching, and indeed learners and teachers, is a complex one and worthy of deeper exploration. Mortimore (1999, p. 3) explains teaching to be 'any conscious activity by one person designed to enhance the learning in another'. One might question the consciousness requirement, as skillful experienced practitioners in the clinical setting, with vast amounts of tacit knowledge, may well subconsciously engage the student in learning through careful role modeling, critical thinking, and problem-solving. These and other methods will be explored as this chapter progresses. Linked to these important approaches to learning is the role of 'teachable moments' in the clinical setting. These are also informal but transformative opportunities for learning. The concept originally gained popularity in 1953 through the work of Havighurst. Teachable moments are spontaneous episodes in the clinical setting that give rise to a practice-based question or uncertainty, which the practitioner explores through reflection and critical thinking, informing their decision-making and problem-solving (Thomas, 2015), thus providing a learning opportunity. For clinical radiographers it is essential to grasp the opportunities for 'teachable moments' to provide transformative, student learning. The concept will be further explored later within this chapter.

Pedagogy is not just the 'science of teaching' but a craft that embraces the student in a more holistic approach; it is not 'done to' the student or learner. Thus, it can be transformative, making changes within the learner who develops a new way of seeing the world (Phillippi, 2010). The exciting yet complex phenomenon of learning will now be further considered.

BLOOM'S TAXONOMY

Learning was categorized in the 1950s by Doctor Benjamin Bloom, an Educational Psychologist, with a view to promoting deeper learning in preference to mere factual recall, or rote learning. The taxonomy was later updated to result in three domains of educational or learning activities, including skill based learning and self-awareness:

- Cognitive or mental, knowledge skills;
- Affective growth in emotions, feeling, and the self;
- Acquisition of psychomotor, manual, or physical skills.

COGNITIVE DOMAIN

Regarding cognitive skills, Bloom's Taxonomy or classification has been adopted worldwide and can be used to structure education, learning, and pedagogy in a range of settings. The higher levels of thinking or metacognition Bloom refers to, enable learners to engage in the analysis and evaluation of complex processes, concepts, procedures, and principles. Such skills are desirable in both radiography practitioners and students. Despite this, a much quoted real-life issue is often that in both the academic and clinical settings: students strategically learn by rote rather than developing deep professional knowledge, questioning, analyzing, and owning their learning. So how might clinical (and academic) radiographers help develop the higher order skills? Potential applications of the Cognitive Domain of Bloom's Taxonomy to radiography are demonstrated in Figure 13.1.

It is generally accepted that skills develop from the lower levels and gradually build towards higher levels of knowledge such as evaluating and creating. So in the clinical setting, students may well have begun 'remembering' and 'understanding' concepts from their academic placements, they must now be encouraged to begin advancing up the pyramid of higher cognition. Thus, it is important to actively question and challenge student learning, aiding their development of the essential critique and analysis skills. Targeted questioning and research rather than blind acceptance should be encouraged. However, this requires practitioners to open themselves up to potentially challenging professional discussions, maybe highlighting areas in which their knowledge could be enhanced. This must not be seen as a reason to shy away from professional challenge and discussion, more the opportunity to create an open questioning culture that drives practice forwards. The cognitive domain is just one aspect of the taxonomy, the affective and psychomotor domains will now be explored.

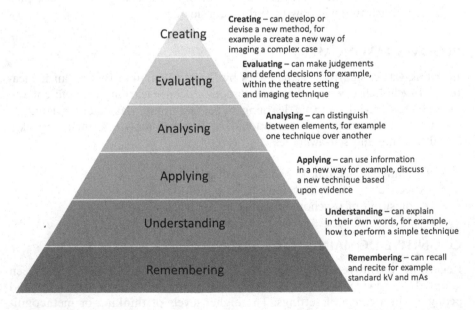

FIGURE 13.1 Cognitive Domain of Bloom's Taxonomy with Suggested Applications to Radiography.

AFFECTIVE DOMAIN

The affective domain (Krathwohl, Bloom, & Masia, 1973) considers the emotional aspects of learning and thus radiography practice. Arguably as important as the cognitive domain, or maybe even more critical, is the manner in which we experience, manage, and learn from feelings, attributes, emotions, values, and motivators. A caring profession such as radiography must be actively influenced by this domain of learning, which should underpin all interactions and opportunities. The phrase 'resilience' is often used in relation to practitioner and student attributes in radiography, helping to avoid the omnipresent 'burn-out' in the highly stressful radiography and radiotherapy workplace (Probst et al., 2012). Rather than 'teaching resilience' *per se*, might it be useful for educators and practitioners to return to the affective domain and consider these categories, from the simple 'receiving phenomena' to the more complex 'internalises values'? This domain may help us develop students into professionals, demonstrating values-based practice in a non-judgmental, unbiased manner. These skills, attributes, feelings and emotions may best be grown in our students through role-modelling and possible patient and public engagement or service user stories, both to be discussed as part of 'learning strategies'.

Figure 13.2 demonstrates aspects of the Affective Domain of Bloom's Taxonomy.

The values-based aspects of practice have recently been published and endorsed by the Society and College of Radiographers (SCoR) (Strudwick et al., 2018). The Association of Radiography Educators considers the concept and

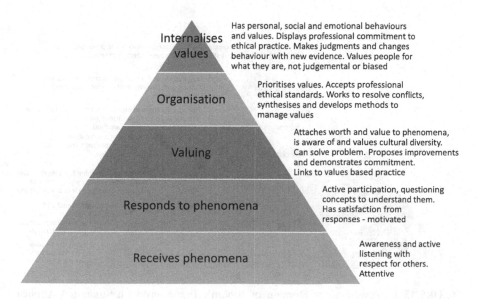

FIGURE 13.2 Affective Domain of Bloom's Taxonomy with Suggested Applications to Radiography.

provides a training handbook to enable practitioners (and students) within both diagnostic and therapeutic radiography to explore the nature of values-based practice (ibid). *'Values-based practice (VBP) is a sister framework to evidence-based practice. Based on learnable clinical skills VBP supports health care professionals to practice shared evidence-based decision-making with their patients, using dialogue about values'* (Strudwick et al., 2018, p. 7). The SCoR have details of VBP and suggested learning opportunities for both practitioners and students. It is beyond the scope of this work to consider VBP in any greater depth but it is an essential resource for learning in radiography practice.

PSYCHOMOTOR DOMAIN

The final, psychomotor, domain was theorized by Simpson (1972) and relates to the development of physical motor skills, relating to movement and co-ordination. These are skills that develop with and through practice, showing enhanced levels of precision, speed and eloquence in the execution of tasks. In radiography, well-coordinated task execution can range from the simple manipulation of basic X-ray equipment through to operating units with more complexity; such as magnetic resonance and ultrasound equipment. Figure 13.3 demonstrates aspects of the Psychomotor Domain of Bloom's Taxonomy.

An understanding of the transition from 'perception' to 'adaptation' is useful for us to know so that students may be supported in their learning. Practice is a key element of this learning and it is essential that adequate time and resource is available for it to be facilitated. Equally, for clinical staff it is important to be aware of the learning process so that expectations of student

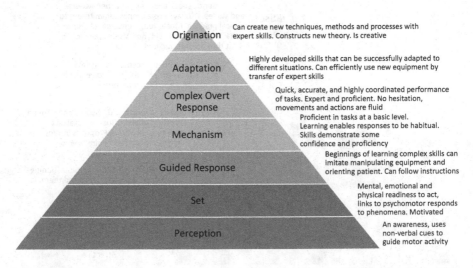

FIGURE 13.3 Psychomotor Domain of Bloom's Taxonomy with Suggested Applications to Radiography.

ability are realistic. Constructive feedback appropriate to the level of student experience can be motivating, another key aspect of learning. As practitioners, this domain of learning can equally apply as one moves from preceptorship through to advanced practice, with skills becoming honed, fluid, and expert.

A brief summary of the stages in 'teaching a skill' to student radiographers is offered in Figure 13.4. This draws together some of the points discussed from Bloom's Taxonomy, and offers a simple practical approach. In this example, competency refers to the specific set of clinical skills, which include the appropriate professional knowledge and attributes. Elements of feedback and feedforward will be explored further within the forthcoming section on barriers and facilitators.

TEACHABLE MOMENTS

With an understanding of the processes involved in student learning, it relies on the practitioner (and student to a degree) to plan learning opportunities and facilitate the practice needed to move from novice to expert. However, the nature of clinical practice is such that unplanned learning opportunities may frequently arise. Despite the clear mandate on responsibility to the patient being foremost in the radiographer role, it is important we do not overlook the prospect of events that might be used to support student learning. These can be defined as 'teachable moments': '... *occasions that spontaneously arise out of a "ready-to-hand" clinical situation in which a practice question, concern, doubt, or uncertainty suddenly surfaces*' (Thomas, 2015, p. 1). Knowledge of learning and how to facilitate this in students can help the use of 'teachable moments' to examine decision-making, explore assumptions and critique actions, thus developing the student and aiding their transition from the lower levels within the taxonomy to the higher order levels. As experienced practitioners, much of the

Demonstrates the skill in its entirety (modelled)
accompanied with step by step clear commentary

↓

Break down the skill into component parts
demonstrate, explain and analyse actions
how each part fits with others must be explained

↓

Skill acquisition supervised, reinforced and practised by students
only by experiencing and repeating can lead to success

↓

Continuous, accurate and swift feedback

↓

Assess skill in parts or as a whole regularly in typical work conditions

FIGURE 13.4   How Can I Teach a Skill or Competency?.

decision-making, skill adaptation, and creativity utilized to solve clinical problems or complications can appear intuitive, and based on tacit knowledge. This 'expert opinion' or synthesis of formal knowledge and clinical expertise can be carefully 'unpicked' by the practitioner, not only for their own reflection and learning, but as a teachable moment. Teachable moments can be facilitated by, and lead to, experiential learning, reflective practice (Schön, 1991), and transformative learning (Mezirow, 1997). We may be more familiar with reflective practice and experiential learning through our own learning endeavors, whilst transformative learning or the development of skills for ongoing autonomous thinking (Mezirow, 1997), may be less familiar. Further reading of texts by Schön and Mezirow is recommended, being beyond the scope of this chapter.

Teachable moments are opportunities for expert, experienced practitioners to 'think aloud', talking through decision-making and exploring the intricacies of their actions. The consideration of their critical thinking and reflection in action (Schön, 1991) externalizes the higher order cognitive domain thinking of Bloom's Taxonomy to the student for evaluation. Arguably it also exposes this thinking to the practitioners themselves, and thus both student and radiographer can learn from the teachable moment. When the practitioner uses the model of thinking aloud, the possibility exists to develop students' theoretical or experiential knowledge towards the well reflected upon, carefully articulated knowing-in-action or tacit, expert knowledge Schön (1991) describes.

Real-life clinical situations are opportunities to apply theory into practice, for students to move from the lower order levels of the cognitive and affective domains towards higher order skills. The complex clinical situation with a real patient interaction offers a special uniqueness that a didactic planned theory session could not convey. The decisions concerning, for example creating the optimal technique when standard positioning is impossible within the accident and emergency setting, are ideal to explore in a teachable moment. Reflecting upon thoughts and actions, and discussing these with the student can allow for a rich learning experience. This experience is centered on the individual person or patient, their unique situation and needs, plus the real time thoughts, processes, and deliberations that have led to the expert decision-making. Referring back to Bloom's Taxonomy, some aspects to consider when using a 'teachable moment' are:

- What cues are drawn upon by the expert as part of Affective – internalizing values to enable professional judgement to be formed.
- What alternatives or options could be utilized in the situation as part of the Cognitive domain – analyzing, evaluating and creating decisions, actions and problem solving.
- What tasks and performance of skills are utilized as part of the Psychomotor domain – adaptation and origination.
- How are tasks performed in an expert, fluid, and proficient manner to be quick, accurate, and highly coordinated, without hesitation, as part of the Psychomotor domain – adaptation and origination.

- How are tasks, thoughts, and actions evaluated in real time by the practitioner and the student.
- When and how judgements are made to adapt a technique, or indeed abandon the imaging all together if the patient, their values and the real life situation indicate.

Ideally as the teachable moment draws to an end, or indeed sometime after, the full circle of potentially transformative learning can be closed. An individual reflection by both student and practitioner, shared with each other and enhanced by open dialogue, can further explore the decisions, practice impacts, patient-specific influences, and a critical evaluation of the unique event and the subsequent learning. Both parties can consider the essential aspect of learning, teaching, and radiography practice: *what could I do differently if this happened again?* Reflection is long established in improving patient care (Lutz et al., 2013), and is a requirement of registered radiographers to reflect upon their practice (HCPC, 2016). It can thus be considered an essential learning strategy whether or not linked to teachable moments.

FURTHER LEARNING STRATEGIES

As previously highlighted, reflection and reflective practice is much espoused by Schön (1991) who describes it as 'reflection in action' and 'reflection on action'. There are essential differences, yet both can be a strategy for learning as individual students or when facilitated by qualified staff. Reflection in action is to reflect on behavior as it happens, whereas, reflection on action reflecting after the event, to review, analyze, and evaluate the situation. These were explored earlier as part of our discussion around teachable moments. Schön also talks about the *'knowing in action'* or the tacit, expert knowledge practitioners may possess, which when shared aloud with students enables teachable moments.

Although reflection facilitates a wide range of learning opportunities to students (and practitioners), including critique of technical and cognitive skills, it may be particularly valuable when considering the softer skills and attributes required of radiographers. Such values or attributes include those highlighted by the NHS Constitution (2019) and include patient-centered care, compassion, dignity and respect, professionalism, and equality. The skills and attributes of the professional radiographer are also described by the Society of Radiographers (2013a) Educational and Career Framework, the Society of Radiographers (2013b) Code of Professional Conduct and the HCPC Standards of Proficiency (2016). These principles will not be further debated here but it is suggested that such values and practices, described by our professional and regulatory bodies, would be best learnt in the clinical setting. The academic environment does have a role in developing these attributes within student radiographers, but curricula may favor a more traditional, behavioristic, pedagogical approach concentrating upon competency, outcomes, and information transfer (Ironside, 2001). Moreover it can be argued that the careful modelling of the 'softer skills' by qualified staff offers the optimal pedagogic method for students to emulate the enactment of core NHS values (Adam & Taylor, 2014). Skillful student reflection, if facilitated by clinical radiographers and linked to teachable moments, can improve the development and enhancement of core values. Arguably, the greatest importance within any interaction is that from the patients' perspective. Bray et al. (2014) suggest the *'human experiences of healthcare'*, when explored by students through interactions and stories, place patients' actual health problems and experiences at the heart of learning.

In a small-scale study involving radiography service users, Strudwick and Harvey-Lloyd (2013) found students benefitted from service user involvement by reflecting upon their experiences and challenging preconceptions. Students identified development of compassionate attributes such as empathy and patient-connectedness. Every patient interaction in radiography could thus be considered a learning opportunity or teachable moment, as every individual will have slightly different needs. It must be reiterated that teachable moments do not only relate to the technical aspects of practice; indeed students report deeper levels of understanding and learning from patients (Flood, Wilson, & Cathcart, 2018). Interactions with patients better enable the development of empathy, compassion, and other professional skills and attributes required of registered radiographers. Yet it must be acknowledged that although the clinical radiography environment can provide immense learning opportunities for students facilitated by staff, it can also be a barrier.

LEARNING AND TEACHING BARRIERS AND FACILITATORS

Within the current NHS, under pressure and under-funded, and despite the policy priority of patient-centered care, there is arguably a greater emphasis upon output, targets and the process-driven business aspects of health (King's Fund, 2014). This could be potentially problematic, creating tensions with an increased emphasis upon technology, diminishing the importance of the 'softer' skills of caring and compassion within the curricula and creating a barrier to student learning (Bolderston, Lewis, & Chai, 2010).

The motivation of typical Generation Z students has been briefly discussed at the start of this chapter. Particular methods or styles of teaching may serve as barriers and facilitators to learning for these students. This generation of student was born with technology; they do not consider it to be an addition to, but more an essential aspect of their life. Generation Z 'lives and breathes' technology, and will readily use it for ten hours daily (Cilliers, 2017). Their attention span is around eight seconds and students demand instant gratification (Kalkhurst, 2018). The brains of Generation Z are more attuned to sophisticated, complex visual imagery so students prefer visual forms of learning, and interaction over communication. In terms of the optimal learning environment, we need to be aware of these preferences to maximize engagement and motivation to learn.

Learning interactions with Generation Z students, who will form the majority of the current student body, should be carefully considered by clinical staff and where possible include the elements of technology, social media and interactions, all of which they find to be so compelling. The technology of modern day radiography is insufficient to fuel the minds of this Generation. Use of visual teaching, active participation, and student-centered learning can create an environment more conducive to thinking and learning. Balancing service demands with staff shortages may not make this an easy task and indeed can act as barriers, yet effective student teaching remains an important aspect of our professional practice.

One of the most important resources to enable student learning is you, the radiographer reading this chapter. You will no doubt recall learning in the clinical setting as a student yourself, the radiographers with whom you could excel, shine, and demonstrate your knowledge whilst being keen and engaged in learning. Conversely, there will be less pleasant memories of the radiographers who destroyed your confidence, motivation, and passion. So how can we be the former, not the latter? This may well be distant role-modelling in action; qualified radiographers might reflect now on their time as a student, characteristics will spring to mind. Many studies examine 'what makes a good facilitator of learning'. For example, the London Deanery Facilitating Learning: Teaching and Learning Paper (McKimm & Jollie, 2007) may be particularly useful, but as a student, practitioner, and educator I would like to suggest the following attributes as depicted in Figure 13.5, which may help fulfill sound facilitator learning.

Through our exploration of 'teachable moments', we highlighted the importance of feedback, discussion, and the need for reflection on learning. Giving students comments, advice and opinions of their learning in a positive, but honest way provides constructive 'feedback' on experiences, performance, actions, and learning opportunities. For students to improve and refine their skills and knowledge they need awareness of 'what went well' and 'what might need refining'. Feedback is a term commonly utilized but ideally, the commentary will feed into or 'feedforward' into future practice and learning. 'Feedforward'

- Be patient, open-minded, and approachable,
- Be enthusiastic and effective,
- Have a good knowledge base, be competent,
- Demonstrate good communication skills, including listening skills,
- Provide encouragement, demonstrates concern, compassion and empathy,
- Take an interest in students as individuals,
- Be aware of the complexities of student life,
- Provide support, be tactful, fair, versatile, confident, and honest,
- Be a good role model, who is non-judgemental,
- Allow for and make time for students and their learning and teaching
- Stimulate thinking
- Provide regular and informative feedback and feedforward,
- Be willing to enhance student confidence and self-esteem,
- Plan and prepare for learning, but also grasp opportunities for 'teachable moments',
- Confirm realistic expectations,

FIGURE 13.5 Attributes Enhancing Good Facilitation of Learning.

enables students to improve their learning and subsequent performance at their next attempt; with the next patient or during the next working day. This crucial part of the learning and teaching process needs to be considerate but be clear and constructive in how to improve. As facilitators we must identify clear actions that will enable improvement, enhancement, or correction. We cannot presume students know what to improve or change, it is essential they are clearly told 'how' and 'what' is required. Balance commentary with motivating positives, but honesty with a considerate, supportive approach is essential. Figure 13.6 suggests some feedback/feedforward essentials, and setting expectations is a good place to begin.

As this chapter draws to a close, it is important to consider the potential of 'failing to fail' students who may not be progressing. Radiographers report some difficulties in failing students as part of clinical assessments or when writing placement reports. The nursing professions (Duffy, 2003) indicate a similar problem; practitioners find an emotional burden when students fail

Ensure clear expectations:

- Communication is essential,
- Students need to take ownership and responsibility for their learning but this is on a continuum from novice to competent,
- Allow students to have control of and thus 'own' their learning,
- Create a culture of inclusion, welcoming and communicating roles and expectations,
- Be objective and reasonable,
- Ensure discussions acknowledge students as adult learners,
- Agree expectations and document

Giving feedback/feedforward:

- Ensure privacy and adequate time for feedback,
- Give students the 'feedback sandwich' (positive/improvement/positive),
- Be constructive – explain what would make it better next time,
- Be sensitive to feelings and student self-esteem,
- Be honest, clear and concise BUT never destructive,
- Be objective, do not use personal remarks but base comments upon skill development of professional attributes,
- Empower and motivate,
- Acknowledge what went well,
- Suggest how and what might correct inappropriate practice,
- Allow comment and clarification from the student,
- Remain calm and objective, seek support if needed,
- Ensure feedback from the team is consistent,
- Generate action plans to help students improve,

FIGURE 13.6   Tips for Feedback/Feedforward.

assessments. They doubt their role as competent teachers and mentors, plus question their caring attributes if they fail students. However we all have a duty of care to the public, the professions and to the students themselves. Failing to fail students with questionable levels of competence may be a result of inadequate action planning and support by academic and clinical educators to enable progression. Early intervention can only be beneficial. But it must be acknowledged that some students do need to fail, that they may never reach the required safe levels of competence, and radiographers must not feel themselves a failure if this is the case. Some tips around managing and supporting students failing to progress can be found in Figure 13.7. For the majority of students, with appropriate support and action planning, there will be progression and success rather than failure.

SUMMARY

Pressures within the clinical radiography setting can make effective student learning and support difficult to achieve on a daily basis. Nevertheless, our professional and regulatory bodies expect the radiographers of the future to be 'grown' by current qualified practitioners in the academic and clinical setting. Despite NHS targets and service demands, student learning, teaching and progression remains a professional requirement. This short chapter on pedagogy, the learning and teaching underpinning radiography practice, has discussed key principles around the process of learning. It is anticipated that with an improved understanding of how learning happens, radiographers in the clinical setting might feel more informed and empowered to facilitate

- If progress is not as expected raise with the student in a supportive way, at the earliest opportunity,
- Do not wait until assessments are due, early intervention is better,
- Explore any obstacles or barriers to learning with the student
- Open communication with the university team at an early stage,
- Formulate a realistic action plan with additional targeted support,
- Carefully document all stages of the process,
- Feed forward regularly to aid progress,
- Celebrate any positives to improve motivation,
- Explore team dynamics and personality issues that may be barriers to student progression,
- Use a dedicated mentor or peer support for the student,
- Early intervention with support from the academic institution can often give result in a positive action plan to support with a successful outcome
- Remember we have a professional duty NOT to pass students, when a fair opportunity to move forward has been offered and they have not reached expectations,

FIGURE 13.7   Managing and Supporting a Student Who Is Not Progressing.

student learning and so improve the students' clinical experience. Subsequently, more students may be successfully completing their education and entering the radiography workforce of the future. In addition, enhanced pedagogic knowledge could be useful to qualified staff and their own learning and progression within the ever-changing demands of NHS practice.

REFERENCES

Adam, D. & Taylor, R. (2014) Compassionate care: Empowering students through nurse education. *Nurse Education Today* 34, 1242–1245.
Augar, P. (2019) Independent panel report to the review of post-18 education and funding. Available at: https://assets.publishing.service.gov.uk/government/uploads/system/uploads/attachment_data/file/805127/Review_of_post_18_education_and_funding.pdf
Bolderston, A., Lewis, D. & Chai, M. (2010) The concept of caring: Perceptions of radiation therapists. *Radiography* 16, 198–208.
Bray, L., O'Brien, M. R., Kirton, J., Zubairu, K. & Christiansen, A. (2014) The role of professional education in developing compassionate practitioners: A mixed methods study exploring the perceptions of health professionals and pre-registration students. *Nurse Education Today* 34, 480–486.
Cilliers, E. J. (2017) The challenge of teaching generation Z. *PEOPLE: International Journal of Social Sciences* 3(1), 188–198.
Duffy, K. (2003) Failing students: A qualitative study of factors that influence the decisions regarding assessment of students' competence in practice. Available at: http://citeseerx.ist.psu.edu/viewdoc/download?doi=10.1.1.515.2467&rep=rep1&type=pdf
Flood, T., Wilson, I. M. & Cathcart, J. (2018) Service user involvement in radiotherapy and oncology education; the patient perspective. *Radiography* 24, 185–191.
Francis, R. (2013) The Mid Staffordshire NHS Foundation Trust public inquiry report. Available at: webarchive.nationalarchives.gov.uk/20150407084231/www.midstaffspublicinquiry.com/report
HCPC. (2016) *Standards of Proficency*. Radiographers. London: Health and Care Professions Council.
Health Education England. (2015) Mind the gap report. Available at: www.nhsemployers.org/-/media/Employers/Documents/Plan/Mind-the-Gap-Smaller.pdf
Ironside, P. (2001) Creating a research base for nursing education: An interpretive review of conventional, critical, feminist, postmodern, and phenomenologic pedagogies. *Advances in Nursing Science* 23(3), 72–87.
Kalkhurst, D. (2018) Engaging Gen Z students and learners. Available at: www.pearsoned.com/engaging-gen-z-students/
King's Fund. (2014) *Developing Collective Leadership for Healthcare*. London. Available at: https://www.kingsfund.org.uk/sites/default/files/field/field_publication_file/developing-collective-leadership-kingsfund-may14.pdf
King's Fund. (2019) Closing the gap report. Available at: www.kingsfund.org.uk/sites/default/files/2019-03/closing-the-gap-health-care-workforce-full-report.pdf
Krathwohl, D. R., Bloom, B. S. & Masia, B. B. (1973) *Taxonomy of Educational Objectives, the Classification of Educational Goals. Handbook II: Affective Domain.* New York: David McKay Co., Inc.
Lexico.com. (2019). https://www.lexico.com/definition/pedagogy.
Lutz, G., Scheffer, C., Edelhaeuser, F., Tauschel, D. & Neumann, M. (2013) A reflective practice intervention for professional development, reduced stress and improved patient care – A qualitative developmental evaluation. *Patient Education and Counselling* 92, 337–345.

McKimm, J. & Jollie, C. (2007) Facilitating learning: Teaching and learning methods. Available at: https://faculty.londondeanery.ac.uk/e-learning/small-group-teaching/Facilitating_learning_teaching_-_learning_methods.pdf

Mezirow, J. (1997) Transformative learning: Theory to practice. *New Directions for Adult and Continuing Education* 74, 5–12.

Mortimore, P. (1999) *Understanding Pedagogy and Its Impact on Learning.* Sage Publications Ltd., London.

NHS Constitution. (2019) Department of Health. Available at: www.gov.uk/government/publications/the-nhs-constitution-for-england/the-nhs-constitution-for-england

Phillippi, J. (2010) Theory-to-practice. Transformative learning in healthcare. *PAACE Journal of Lifelong Learning* 19, 39–54.

Probst, H., Griffiths, S., Adams, R. & Hill, C. (2012) Burnout in therapy radiographers in the UK. *The British Journal of Radiology* 85, e760–e765.

Schön, D. A. (1991) *The Reflective Practitioner: How Professionals Think in Action.* Taylor & Francis Ltd., England.

Simpson, E. J. (1972) *The Classification of Educational Objectives in the Psychomotor Domain.* Washington, DC: Gryphon House.

Society of Radiographers. (2013b) Code of professional conduct. Available at: www.sor.org/learning/document-library/code-professional-conduct

Society of Radiographers. (2013a) Educational and career framework for the radiography workforce. Available at: www.sor.org/learning/document-library/download-formats/mobi/8938

Strudwick, R. & Harvey-Lloyd, J. (2013) Preparation for practice through service user involvement in the diagnostic radiography curriculum at University Campus Suffolk. *PBLH*, The Higher Education Academy. 1, 2.

Strudwick, R., Newton-Hughes, A., Gibson, S., Harris, J., Gradwell, M., Hyde, E., Harvey-Lloyd, J., O'Regan, T. & Hendry, J. (2018) *Values-Based Practice (VBP) in Diagnostic and Therapeutic Radiography. A Training Template.* College of Radiographers and the Collaborating Centre for Values-Based Practice in Health and Social Care. Available at: www.sor.org/sites/default/files/document-versions/2018.10.03_radiography_vbp_training_manual_-_final.pdf

Thomas, L. (2015) *Critical Thinking and Clinical Reasoning in the Health Sciences.* PA Facione and NC Facione (eds.). California Academic Press. Available at: www.insightassessment.com/About-Us/Measured-Reasons/pdf-file/aa-Clinical-Reasoning-Resource-PDFs/Teachable-Moments-in-Clinical-Practice-Promoting-Clinical-Judgment/(language)/eng-US [Accessed July 2019].

Wilkinson, E. (2015) UK NHS staff: Stressed, exhausted, burnt out. *The Lancet* 385, 841–842.

Index

Please note: Page numbers with f indicate figures and those with t indicate tables.

Printed in the United States
by Baker & Taylor Publisher Services